buchpark
AUSGESONDERT
www.buchpark.de

Ernst Schering Research Foundation Workshop 57
The Histone Code and Beyond

Ernst Schering Research Foundation
Workshop 57

The Histone Code and Beyond

New Approaches to Cancer Therapy

S.L. Berger, O. Nakanishi, B. Haendler
Editors

With 36 Figures

Series Editors: G. Stock and M. Lessl

Library of Congress Control Number: 2005932740

ISSN 0947-6075

ISBN-10 3-540-27857-5 Springer Berlin Heidelberg New York
ISBN-13 978-3-540-27857-3 Springer Berlin Heidelberg New York

Springer is a part of Springer Science+Business Media
springeronline.com

Printed in Germany

Editor: Dr. Ute Heilmann, Heidelberg
Desk Editor: Wilma McHugh, Heidelberg
Production Editor: Monika Riepl, Leipzig
Cover design: design & production, Heidelberg
Typesetting and production: LE-TeX Jelonek, Schmidt & Vöckler GbR, Leipzig
21/3150/YL – 5 4 3 2 1 0 Printed on acid-free paper

Preface

In recent years, it has been recognized that epigenetic marks play an essential role in the control of gene expression by modulating the access of regulatory factors to chromatin. Evidence is now accumulating that deregulation of these control mechanisms is linked to cancer initiation and progression.

In the chromatin, DNA is packaged around histone proteins and generally in the default state of repression. DNA regions that are to be transcribed are temporarily made accessible by decondensing the structure. For the uncoiling of chromatin, histones can be post-translationally modified at many positions, mainly located in their N-terminal tails. A range of histone modifications including phosphorylation, acetylation, methylation and ubiquitylation, all carried out by different families of specific enzymes, have been identified in recent years.

Another epigenetic event is DNA methylation, which often takes place on cytosine residues in CpG islands found in gene promoter regions. Aberration in DNA methylation is now recognized as a fundamental aspect of human cancer, each tumor type having its own DNA methylation pattern.

Besides histones, post-translational modifications affect and regulate a variety of additional proteins with important functions in signal transduction and gene transcription. Ubiquitylation, sumoylation and neddylation represent three related processes that may dramatically affect the function of targeted proteins, for instance by regulating their stability, subcellular localization or interactions with other proteins.

All these exciting developments in the area of chromatin research prompted us to organize a meeting to discuss the most recent breakthroughs made in this fast-moving field, with a focus on aspects relevant to cancer research. The following topics were selected for the workshop: chromatin dynamics, histone and DNA methylation, histone code and covalent modifications of nonhistone proteins.

We were pleased to be able to bring together internationally highly recognized experts who study this field using a variety of biochemical, genetic and structural approaches. We are grateful to all of them for their willingness to participate and for their excellent contributions.

We hope that the proceedings of this workshop will promote a better understanding of the mechanisms responsible for epigenetic control of gene expression and help to identify novel targets for tumor treatment in the near future.

Shelley L. Berger
Osamu Nakanishi
Bernard Haendler

Contents

List of Editors and Contributors

Editors

Berger, S.L.
The Wistar Institute, Gene Expression and Regulation Program
3601 Spruce Street, Room 201, PA 19104 Philadelphia, USA
(e-mail: berger@wistar.org)

Nakanishi, O.
CRBA Oncology, Nihon Schering, Kobe, Japan
(e-mail: onakanishi@schering.com.jp)

Haendler, B.
CRBA Oncology, Schering AG,
Müllerstr. 178, 13342 Berlin, Germany
(e-mail: bernard.haendler@schering.de)

Contributors

Bao, Y.
Department of Biochemistry and Molecular Biology,
Colorado State University,
Fort Collins, CO 80523-1870, USA

Caron, C.
Laboratoire de Biologie Moléculaire et Cellulaire de la Différenciation,
INSERM U 309, Equipe Chromatine et Expression des Gènes,
Institut Albert Bonniot, Faculté de Médecine, Domaine de la Merci,
38706 La Tronche Cedex, France

Chodaparambil, J.Y.
Department of Biochemistry and Molecular Biology,
Colorado State University,
Fort Collins, CO 80523-1870, USA

Edayathumangalam, R.S.
Department of Biochemistry and Molecular Biology,
Colorado State University,
Fort Collins, CO 80523-1870, USA

Emre N.C.T.
Gene Expression & Regulation Program, The Wistar Insitute,
3601 Spruce St. Rm 201, Philadelphia, PA 19104, USA

Esteller, M.
Spanish National Cancer Center (CNIO) Epigenetics Laboratory,
Molecular Pathology Programme,
Melchor Fernandez Almagro 3, 28029 Madrid, Spain
(e-mail: mesteller@cnio.es)

Fodor, B.D.
Research Institute of Molecular Pathology (IMP), The Vienna Biocenter,
Dr. Bohrgasse 7, 1030 Vienna, Austria
(e-mail:fodor@imp.univie.ac.at)

Govin, J.
Laboratoire de Biologie Moléculaire et Cellulaire de la Différenciation,
INSERM U 309, Equipe Chromatine et Expression des Gènes,
Institut Albert Bonniot, Faculté de Médecine, Domaine de la Merci,
38706 La Tronche Cedex, France

Hay, R.T.
Centre for Biomolecular Sciences, School of Biology,
University of St. Andrews, Building North Haugh,
St. Andrews, KY 169ST Fife, Scotland, UK
(e-mail: rth@st-and.ac.uk)

Jenuwein, Th.
Research Institute of Molecular Pathology (IMP), The Vienna Biocenter,
Dr. Bohrgasse 7, 1030 Vienna, Austria
(e-mail: jenuwein@imp.univie.ac.at)

Khochbin, S.
Laboratoire de Biologie Moléculaire et Cellulaire de la Différenciation,
INSERM U 309, Equipe Chromatine et Expression des Gènes,
Institut Albert Bonniot, Faculté de Médecine, Domaine de la Merci,
38706 La Tronche Cedex, France
(e-mail: khochbin@ujf-grenoble.fr)

Kohlmaier, A.
Research Institute of Molecular Pathology (IMP), The Vienna Biocenter,
Dr. Bohrgasse 7, 1030 Vienna, Austria
(e-mail: kohlmaier@imp.univie.ac.at)

Kubicek, St.
Research Institute of Molecular Pathology (IMP), The Vienna Biocenter,
Dr. Bohrgasse 7, 1030 Vienna, Austria
(e-mail: kubicek@imp.univie.ac.at)

Lachner, M.
The Rockefeller University,
1230 York Avenue,New York, NY 10021, USA
(e-mail: mlachner@mail.rockefeller.edu)

Lestrat, C.
Laboratoire de Biologie Moléculaire et Cellulaire de la Différenciation,
INSERM U 309, Equipe Chromatine et Expression des Gènes,
Institut Albert Bonniot, Faculté de Médecine, Domaine de la Merci,
38706 La Tronche Cedex, France

Linderson, Y.
Research Institute of Molecular Pathology (IMP), The Vienna Biocenter,
Dr. Bohrgasse 7, 1030 Vienna, Austria
(e-mail: linderson@imp.univie.ac.at)

Luger, K.
Department of Biochemistriy and Molecular Biology,
Colorado State University,
Fort Collins, CO 80523-1870, USA
(e-mail: kluger@lamar.colostate.edu)

Martens, J.H.A.
Research Institute of Molecular Pathology (IMP), The Vienna Biocenter,
Dr. Bohrgasse 7, 1030 Vienna, Austria
(e-mail: martens@imp.univie.ac.at)

Mellor, J.
Department of Biochemistry,
Oxford, South Parks Road, OX1 3 QU Oxford, UK
(e-mail: jane.mellor@bioch.ox.ac.uk)

Nakatani, Y.
Dana-Farber Cancer Institute and Harvard Medical School,
MA 02115 Boston, USA
(e-mail: yoshihiro_nakatani@dfci.harvard.edu)

O'Sullivan, R.J.
Research Institute of Molecular Pathology (IMP), The Vienna Biocenter,
Dr. Bohrgasse 7, 1030 Vienna, Austria
(e-mail: osullivan@imp.univie.ac.at)

Owen-Hughes, T.
Division of Gene Regulation and Expression, School of Life Sciences,
University of Dundee,
DD1 5EH, Dundee, Scotland UK
(e-mail: t.a.owenhughes@dundee.ac.uk)

Park, Y.-J.
Department of Biochemistry and Molecular Biology,
Colorado State University,
Fort Collins, CO 80523-1870, USA

Perez-Burgos, L.
Research Institute of Molecular Pathology (IMP), The Vienna Biocenter, Dr. Bohrgasse 7, 1030 Vienna, Austria
(e-mail: perez@imp.univie.ac.at)

Peters, A.H.F.M.
Friedrich Miescher Insitute for Biomedical Research (FMI)
Novartis Research Foundation,
Maulbeerstrsse 66, CH-4058 Basel, Switzerland
(e-mail: Antoine.Peters@fmi.ch)

Pirrotta, V.
Department of Molecular Biology and Biochemistry, Rutgers University, 604 Allison Road, Piscataway NJ 08854, USA
(e-mail: pirrotta@biology.rutgers.edu)

Pivot-Pajot, Ch.
Laboratoire de Biologie Moléculaire et Cellulaire de la Différenciation, INSERM U 309, Equipe Chromatine et Expression des Gènes, Institut Albert Bonniot, Faculté de Médecine, Domaine de la Merci, 38706 La Tronche Cedex, France

Rousseaux, S.
Laboratoire de Biologie Moléculaire et Cellulaire de la Différenciation, INSERM U 309, Equipe Chromatine et Expression des Gènes, Institut Albert Bonniot, Faculté de Médecine, Domaine de la Merci, 38706 La Tronche Cedex, France

Schotta, G.
Research Institute of Molecular Pathology (IMP), The Vienna Biocenter, Dr. Bohrgasse 7, 1030 Vienna, Austria
(e-mail: schotta@imp.univie.ac.at)

Sengupta, R.
Research Institute of Molecular Pathology (IMP), The Vienna Biocenter, Dr. Bohrgasse 7, 1030 Vienna, Austria
(e-mail: sengupta@imp.univie.ac.at)

Sharrocks, A.
Faculty of Life Sciences, University of Manchester,
Michael Smith Building,
Oxford Road, Room A2028, M13 9PT Manchester, GB
(e-mail: a.d.sharrocks@manchester.ac.uk)

Shestakova, E.
Dana-Farber Cancer Institute and Harvard Medical School,
MA 02115 Boston, USA

Tagami, H.
Division of Biological Science, Graduate School of Science,
Nagoya University, Chikusa, Nagoya, Aichi 464-8602, Japan

Yang, S.-H.
Faculty of Life Sciences, University of Manchester,
Michael Smith Building,
Oxford Road, Manchester, M13 9PT, UK

Yonezawa , M.
Research Institute of Molecular Pathology (IMP), The Vienna Biocenter,
Dr. Bohrgasse 7, 1030 Vienna, Austria
(e-mail: yonezawa@imp.univie.ac.at)

1 The Role of Histone Modifications in Epigenetic Transitions During Normal and Perturbed Development

S. Kubicek, G. Schotta, M. Lachner, R. Sengupta, A. Kohlmaier, L. Perez-Burgos, Y. Linderson, J.H.A. Martens, R.J. O'Sullivan, B.D. Fodor, M. Yonezawa, A.H.F.M. Peters, T. Jenuwein

Abstract. Epigenetic mechanisms control eukaryotic development beyond DNA-stored information. DNA methylation, histone modifications and variants, nucleosome remodeling and noncoding RNAs all contribute to the dynamic make-up of chromatin under distinct developmental options. In particular, the great diversity of covalent histone tail modifications has been proposed to be ideally suited for imparting epigenetic information. While most of the histone tail modifications represent transient marks at transcriptionally permissive chromatin, some modifications appear more robust at silent chromatin regions, where they index repressive epigenetic states with functions also outside transcriptional regulation. Under-representation of repressive histone marks could be indicative of epigenetic plasticity in stem, young and tumor cells, while committed and senescent (old) cells often display increased levels of these more stable modifications. Here, we discuss profiles of normal and aberrant histone lysine methylation patterns, as they occur during the transition of an embryonic to a differentiated cell or in controlled self-renewal vs pro-neoplastic or metastatic conditions. Elucidating these histone modification patterns promises to have important implications for novel advances in stem cell research, nuclear reprogramming and cancer, and may offer novel targets for the combat of tumor cells, potentially leading to new diagnostic and therapeutic avenues in human biology and disease.

1.1 The Distinction Between Genome and Epigenomes

The DNA double helix is the prime macromolecule to store and propagate genetic information. The key genetic principle is mutation of one or many of the four nucleotides, resulting in a change in the DNA sequence that can ultimately define a new species barrier. The genome of the unicellular eukaryote *Saccharomyces cerevisiae* consists of around 6,000 genes (Goffeau et al. 1996), a number sufficient for controlling basic cellular processes, such as metabolism, cell division, and DNA damage repair. Mammals have five times as many genes (Lander et al. 2001; Waterston et al. 2002), and most of these additional genes can be associated with specialized functions to determine cell type identities and lineage commitment. This results in the phenotypical discrimination of more than 200 cell types. How can these distinct cell types be defined and stably propagated from a common genome?

In eukaryotes, the DNA molecule is not naked, but is organized in chromatin – the dynamic template of the genetic information. The basic repeating unit of chromatin is the nucleosome, consisting of 147 bp of DNA wrapped around an octamer of the core histones H2A, H2B, H3, and H4 (Luger et al. 1997). One of the key epigenetic principles for altering chromatin states are covalent modifications, both of the DNA (methylation) and of the flexible histone N-termini (histone "tails"). An epigenome can thus be defined by the sum of biochemical modifications of the chromatin template. Epigenomes will greatly differ between distinct cell types and among resting vs proliferative cells, and it has been suggested that histone modifications could reflect a histone code that may index diverse developmental or proliferative options (Strahl and Allis 2000; Turner 2000; Jenuwein and Allis 2001). Currently, more than 30 amino acid positions have been described in the N-termini of the four core histones to be subject to distinct post-translational modifications. Although this large number may be counter-intuitive for a histone code and rather reflect biochemical affinities for chromatin-associated factors (Schreiber and Bernstein 2002), several histone modifications, and in particular distinct combinations of selective histone marks, appear to be of predictive value (see Sects. 1.2 and 1.3).

In addition to DNA methylation, histone modifications and variants (Sarma and Reinberg 2005), nucleosome remodeling factors (Narlikar et al. 2002) and noncoding RNAs (Lippman and Martienssen 2004) all cooperate to organize chromatin into accessible (euchromatic) or inaccessible (heterochromatic) subdomains. These epigenetic mechanisms considerably extend the information potential of the genetic code and are important for the differential usage of gene function during cell type specification. Thus, one genome can generate many epigenomes (Fig. 1), as the fertilized egg progresses through development and translates its information into a multitude of cell fates. As a general pattern, activating epigenetic mechanisms appear to prevail in stem, young, and tumor cells, since these cells would allow access to the nearly full gene comple-

Fig. 1. Genome vs epigenomes. The figure illustrates the distinction between genome (DNA sequence) and epigenome (the collective modification pattern of chromatin). Diverse mechanisms including DNA methylation, histone modifications and variants, chromatin-remodeling activities, and noncoding RNAs contribute to the generation of distinct epigenomes, as the fertilized egg progresses during development. The various epigenetic modifications are indicated as histone acetylation (*blue flag*), histone lysine methylation (*green hexagons* for active and *red* or *blue hexagons* for repressive marks), histone arginine methylation (*orange hexagon*) DNA methylation (*small brown hexagons*), histone variants (*yellow nucleosome*) and remodeling complexes (e.g., ISWI). Euchromatin allows transcription factors (TF) to access the underlying DNA sequence, while repressive modifications that may be induced by small noncoding RNAs could lead to the further compaction of heterochromatic domains

ment. A transcriptionally permissive state would therefore be indicative of cellular plasticity and pluripotency. By contrast, repressive mechanisms accumulate in committed and old cells and stabilize developmental options by restricting expression of lineage-inappropriate genes. Both activating and repressive epigenetic mechanisms are important for transcriptional control, but also play major roles in overall chromatin

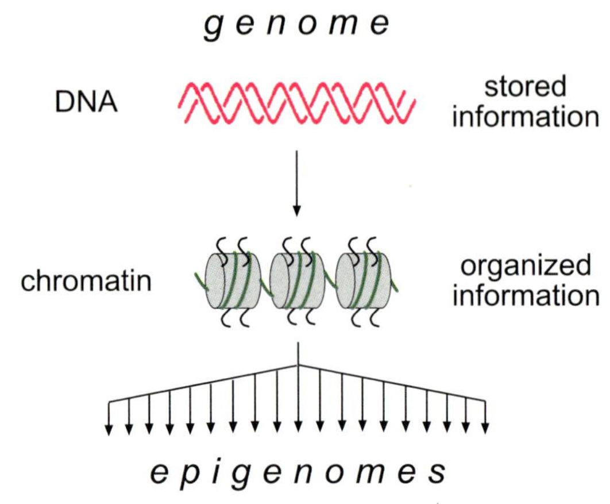

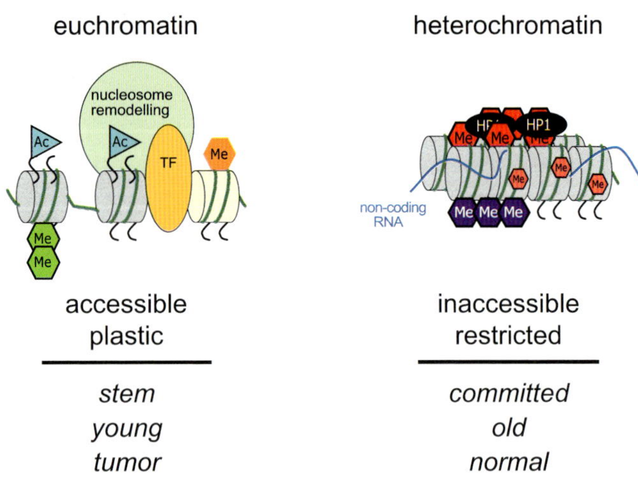

organization to partition eukaryotic chromosomes into centromeric and telomeric domains, to index recombination sites and to respond to DNA damage repair. The implications of epigenetic mechanisms for human biology and disease, including genome stability, cancer, stem cells, and aging are far-reaching.

1.2 Activating Histone Modifications

Histone modifications include acetylation, phosphorylation, methylation (of arginines and lysines), ubiquitination, sumoylation and ADP-ribosylation (van Holde 1988). These modifications are predominantly found in the flexible N-termini of the small (102–135 amino acids) histone proteins. One nucleosome contains two copies of the four core histones, and approximately 30 amino acids (the histone tails) of each histone molecule protrude from one nucleosome. If only these amino acids are considered, this would result in the potential of $2 \times 4 \times 30$ positions conferring epigenetic information per nucleosome. That each amino acid is important is suggested by the extremely high evolutionary conservation of the histone sequences. However, histone modifications do not appear to occur in isolation, but rather in a combinatorial manner as proposed for modification cassettes (Fischle et al. 2003) and trans-histone pathways (Briggs et al. 2002). Intriguingly, almost all of the known histone modifications can either have an activating or a repressive function, dependent on which amino acid position(s) in the histone N-termini are modified (see Sect. 1.3). Both synergistic and antagonistic pathways have been described (Berger 2002) that can progressively induce combinations of active marks while simultaneously counteracting repressive modifications. It is, however, not known how many distinct combinations for modifications of the various N-terminal histone positions indeed exist in any given nucleosome, although some of the major sites can be modified at 60%–80% in bulk histone preparations (Peters et al. 2003; Rice et al. 2003). In addition to the N-termini, modifications in the globular histone fold domains have recently been shown to also impact chromatin structure and assembly, thereby influencing gene expression and DNA damage repair (van Leeuwen et al. 2002; Ng et al. 2002; Freitas et al. 2003; Xu et al. 2005, Ye et al. 2005).

Activating histone tail modifications have been shown to comprise general H3 and H4 acetylation, H3S10 phosphorylation, arginine methylation at H3R17 and lysine methylation at H3K4 and H3K36 positions (Lachner et al. 2003), although there are many others. Histone acetyltransferases (HATs), kinases, protein arginine methyltransferases (PRMTs), and histone methyltransferases (HMTases) are the enzymes that confer these modifications (Fig. 2). The various activating histone modifications are induced by these enzymatic activities, which can either be associated in multi-protein complexes or are recruited to the chro-

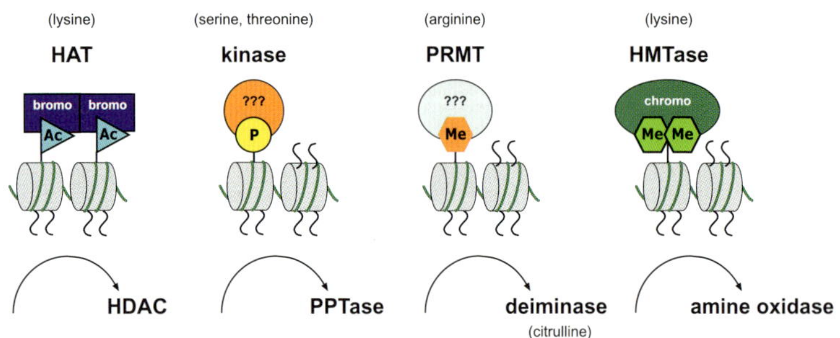

Fig. 2. Activating histone modifications and associated enzymes. Activating histone modifications comprise lysine acetylation (*blue flag*), serine and threonine phosphorylation (*yellow circle*), arginine methylation (*orange hexagon*), and lysine methylation at positions H3K4, H3K36, and H3K79 (*green hexagons*). Lysine acetylation is recognized by bromo-domain factors, while lysine methylation can recruit chromo-domain proteins. Specific binders for histone phosphorylation and arginine methylation are currently not known. Activating modifications are induced by histone acetyltransferases (*HATs*), kinases, protein arginine methyltransferases (*PRMTs*), and histone lysine methyltransferases (*HMTases*). Most of these marks are transient and can be actively removed by histone deacetylases (*HDACs*), protein phosphatases (*PPTases*), deiminases and amine oxidases

matin template in a sequential order by transcription factors, nucleosome remodelers or the RNA-PolII machinery (Sarma and Reinberg 2005). Histone modifications can result in *cis*-effects, which alter electrostatic interactions of the DNA polymer with nucleosomes. For example, both acetylation and phosphorylation will reduce the net positive charge of the basic histone molecules, thereby increasing overall accessibility of nucleosomal DNA. Alternatively, histone modifications can generate affinities for chromatin-associated factors, which then mediate transitions of the chromatin template by *trans*-mechanisms. This has been shown for histone acetylation in providing a recruiting signal for positively acting bromo domain factors (Dhalluin et al. 1999; Jacobson et al. 2000) and for histone lysine methylation, which represents binding sites for chromo domain (Daniel et al. 2005) or WDR5 proteins (Wysocka et al. 2005). Currently, no protein modules that selectively recognize histone arginine methylation or H3S10 phosphorylation are known.

Most activating modifications are transient and can be actively removed, for example by histone deacetylases (HDACs), protein phosphatases (PPTases) and peptidyl arginine deiminases (PADs). PADs convert unmodified or monomethylated arginine to citrulline, thereby inducing an alteration in amino acid composition (Wang et al. 2004; Cuthbert et al. 2004). The function and stability of citrulline in histones or nucleosomes is currently not clear. In contrast to arginine methylation, lysine methylation is considered a chemically very stable modification, as the amino-methyl bond cannot be cleaved directly. However, recently a lysine specific "demethylase" (LSD1) was described as an amine oxidase that is able to remove H3K4 methylation (Shi et al. 2004). The enzyme acts by oxidative destabilization of the amino-methyl bond, resulting in unmodified lysine and formaldehyde. LSD1 was shown to be selective for the activating H3K4 methylation mark and can only destabilize mono- and di-, but not trimethylation. This demethylase is part of a large repressive protein complex that also contains HDACs and other enzymes (Shi et al. 2004). Similar to the collaboration of activating enzymes (e.g., HATs and PRMTs) in inducing combinations of positive histone modifications, there is synergy between antagonizing enzymes (e.g., HDACs and PADs) in removing or counteracting these activating marks.

1.3 Repressive Histone Modifications

In addition to DNA methylation (Jaenisch and Bird 2003), many of the known histone modifications can also exert a repressive function on the chromatin template. In particular, histone lysine methylation has emerged as a central epigenetic modification that can index silenced chromatin regions. There are five well-characterized and prominently methylated lysine positions in the histone H3 and H4 N-termini, although there are additional methylation sites also in the globular domains (van Leeuwen et al. 2002; Ng et al. 2002) and in the linker histone H1 (Kuzmichev et al. 2004). Whereas H3K4 and H3K36 methylation are activating marks (see Sect. 1.2), histone H3 lysine 9 (H3K9), lysine 27 (H3K27) and histone H4 lysine 20 (H4K20) methylation represent OFF marks that are mainly involved in the organization of repressive chromatin structures (Peters et al. 2003; Schotta et al. 2004). All of the five major histone lysine methylation sites can exist in three distinct states: mono-, di- and trimethylation (Fig. 3).

The existence of three lysine methylation states provides an additional layer of regulatory control, since mono-, di- and trimethylation may serve diverse biological functions. For example, there are more than 50 SET-domain HMTases in mammals (Schneider et al. 2002; Schotta et al. 2004), which differ in their potential to induce mono- vs di- vs trimethylation. Dependent on a distributive or processive activity (Zhang et al. 2003), some of these enzymes can only establish monomethylation (e.g., PR-SET7) (Nishioka et al. 2002; Xiao et al. 2005), while others convert an unmodified substrate to several methylation states (e.g., EZH2) (Kuzmichev et al. 2004) or are specialized to induce the fully methylated end state of trimethylation (e.g., Suv39h) (Peters et al. 2003). Methylation at distinct histone lysine positions generates selective binding sites for chromatin-associated factors containing a chromo domain, as was first demonstrated for HP1 to interact with H3K9me3 and for Polycomb to display affinity toward H3K27me3 (Daniel et al. 2005; Ringrose and Paro 2004).

Recently, the chromo-domain superfamily of "royal proteins" (Maurer-Stroh et al. 2003) has been shown to comprise additional methyllysine binders, such as tudor domain (Sanders et al. 2004) or malignant brain tumor (MBT) (W. Fischle and C.D. Allis, personal communication) proteins. Interestingly, some of these factors not only discriminate between distinct methyl-lysine positions in the histone N-termini, but also are selective to recognize a mono-, di- or trimethyl state (Fig. 3). Currently, no enzymes removing repressive histone lysine trimethyl marks are known. This is consistent with the apparent stability of H3K9, H3K27, and H4K20 trimethylation, which can persist through mitosis and over many cell generations (Peters et al. 2003; Schotta et al. 2004). This robustness has been used to suggest that repressive histone lysine

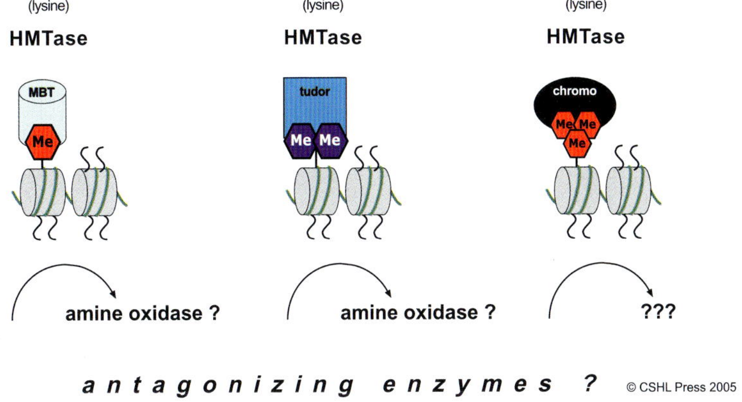

Fig. 3. Repressive histone lysine methylation states. Lysine methylation at H3K9, H3K27, and H4K20 positions are among the best characterized repressive histone modifications. The ε-amino group of lysines can be modified by mono-, di-, and trimethylation. Originally, chromo domain proteins were the first factors identified to bind methylated lysines in the histone N-termini. This group of methyl-lysine binders has recently been extended by members of the royal family, such as malignant brain tumor (*MBT*) and tudor domains. Histone lysine methylation is induced by HMTases that can differ in their potential to generate mono-, di-, or trimethylation. Whereas mono- and dimethyl states can be destabilized by amine oxidases, more potent mechanisms (e.g., radical attack and hydroxylation) are required to remove a trimethyl state. Currently, no enzymes are known that would destabilize repressive histone lysine trimethyl marks

trimethyl states may be particularly suited to impart epigenetic memory to a differentiating cell. However, in analogy to LSD1 described in Sect. 1.2, it is formally possible for an amine oxidase to destabilize repressive histone lysine mono- or dimethyl states. Such enzymes may most likely be present in diverse protein complexes with other activating enzymes, such as HATs or PRMTs and HMTases (e.g., trx and MLL) that can induce positive acetylation and methylation marks. Indeed, an LSD1-like activity to counteract repressive H3K9me2 has recently been described for the transcriptional activation of genes that respond to certain hormones (Metzger et al. 2005). Here, LSD1 can associate with the androgen receptor in a complex formation, which apparently converts the substrate specificity of LSD1 at target promoters to act on H3K9me2 rather than on H3K4me2. Even when complexed with the androgen receptor (which may induce an altered substrate presentation of the H3 N-terminus), LSD1 cannot convert H3K9me3 (Metzger et al. 2005), and the enzymatic reaction of amine oxidation is incompatible to destabilize lysine trimethylation. Alternative mechanisms have been proposed, and attack of the amino-methyl bond by oxygen radicals involving hydroxylation could represent a plausible pathway. In particular, conserved jumonji (jmjC) domain enzymes, which require Fe^{2+} and oxoglutarate as co-factors, have been suggested as potential hydroxylases that could remove histone lysine trimethylation (Trewick et al. 2005). However, since repressive histone lysine trimethyl states appear very stable, such hydroxylating enzymes may be of very low abundance or could be extremely tightly regulated during early development and in lineage commitment. There are ongoing experiments in many laboratories to identify these putative methyl-lysine hydroxylases, which promise to be novel candidates for exploratory research, since their neutralization by small molecule inhibitors would increase chromosome stability and potentially counteract plasticity of gene expression and pro-neoplastic dysfunction.

1.4 Reprogramming of Epigenetic Modifications

Alterations in DNA methylation patterns (Bird 2002), misregulated chromatin remodeling (Klochendler-Yeivin et al. 2002; Narlikar et al. 2002), and changes in histone modifications (Schneider et al. 2002) have been linked with perturbed development, tumorigenesis (Jones and Baylin

2002), and with inefficient reprogramming of cloned mammalian embryos after nuclear transfer (Jaenisch and Bird 2003).

Although cell fate specification in mammals allows for approximately 200 different cell types, there are, in principle, two major transitions: from a stem (pluripotent) cell to a fully differentiated cell and between a resting (quiescent) and a proliferating cell. These represent the extreme endpoints among many intermediates, consistent with a multitude of different make-ups of the epigenome in mammalian development. During embryogenesis, a dynamic increase of epigenetic modifications is detected in the transition from the fertilized oocyte to the blastocyst stage, and then at implantation, gastrulation, organ development and fetal growth. Most of these modifications or imprints may be erasable

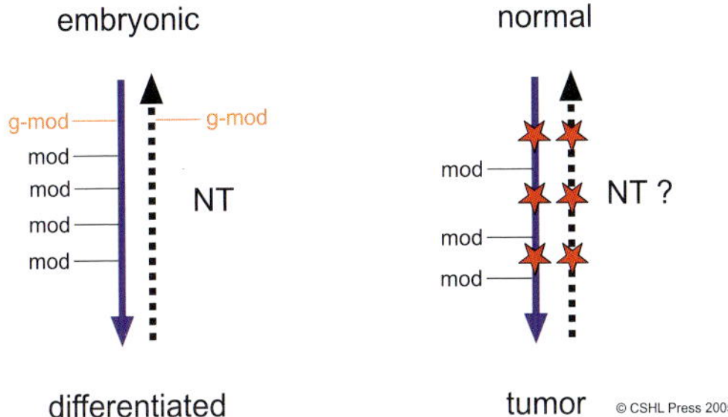

Fig. 4. Overview for reprogramming by nuclear transfer. Differentiation of distinct cell types from a pluripotent embryonic cell is accompanied by the acquisition of epigenetic modifications (*mod*), such as DNA methylation, histone modifications and variants, and chromatin remodeling activities. While most of these modifications can be removed after nuclear transfer (*NT*) to an enucleated oocyte, some germline-modifications (*g-mod*) are retained in the cloned embryo. In contrast, tumorigenesis is invariably triggered by genetic lesions (e.g., mutations, deletions, translocations; *red asterisks*) that are irreversible by NT. In addition, tumor progression is correlated with aberrant epigenetic modifications (different from those gained during normal development), most of which can be reprogrammed by NT. (Modified after R. Jaenisch, 2003; personal communication)

via transfer of a differentiated nucleus to the cytoplasm of an enucleated oocyte. However, some marks may persist, thereby restricting normal development of cloned embryos (Rideout et al. 2001; Jaenisch et al. 2005), and a few could even be inherited as germline modifications (g-mod) (Fig. 4), which, in mammals, are likely to include DNA methylation. Interestingly, in *Schizosaccharomyces pombe*, Swi6-dependent epigenetic variegation can be silenced for many cell divisions both during mitosis and meiosis (Grewal and Klar 1996) by histone modifications (most probably H3K9me2) that are independent of DNA methylation. Analogous studies were performed in *Drosophila* using a pulse of an activating transcription factor to transmit cellular memory for *Hox* gene expression during the female germline (Cavalli and Paro 1999). In both of these examples, epigenetic memory is mediated by chromatin alterations that comprise distinct histone modifications and, most likely, also incorporation of histone variants (Henikoff and Ahmad 2005).

Neoplastic transformation and tumorigenesis can be regarded as misregulated differentiation with perturbed proliferation control. Often, these transformations can be correlated with the accumulation of aberrant epigenetic modifications (different from those gained during normal development) (Fig. 4). Epigenetic dysregulation can alter the tumor profile and the aggressiveness or metastatic potential of cancer cells. To which extent DNA methylation and chromatin modifications contribute to the epigenetic make-up of a given tumor was addressed by nuclear transfer of a cancer cell nucleus and subsequent analysis of the tumor-prone cloned mice (Hochedlinger et al. 2004). Since there is no tumor without primary DNA lesions (e.g., point mutations, deletions or translocations), which are irreversible by nuclear transfer, all cloned mice from the melanoma tumor cells invariably developed cancer. However, their tumor spectrum greatly varied (Hochedlinger et al. 2004), consistent with different contributions of epigenetic modifications to trigger neoplastic progression.

1.5 Persistence of Soma-Specific Epigenetic Marks in Cloned Embryos

DNA methylation and histone modifications are asymmetrically distributed in parental pronuclei, and these differences can be visualized prior to nuclear fusion in the mammalian zygote (Arney et al. 2002; Santos et al. 2003; Morgan et al. 2005). For example, DNA methylation of the paternal genome is rapidly lost after fertilization, and repressive H3K9me2 and H3K9me3 states are absent in the male pronucleus but present in maternal chromatin (Santos et al. 2005). In addition, asymmetry of histone variants (van der Heijden et al. 2005) and transfer of nucleosomes from maternal to paternal DNA can either enhance or balance epigenetic differences of the two parental genomes. One of the most illuminating examples is imprinted X-inactivation, which always occurs on the paternal X chromosome in trophectoderm cells (an extraembryonic tissue that gives rise to the placenta). In contrast, the paternal X chromosome is reactivated in the inner cell mass (the embryo proper) of blastocysts (Okamoto et al. 2004; Mak et al. 2004), which only later enter into random X-inactivation by the time of implantation. Intriguingly, there is a strong correlation between repressive histone lysine methylation marks (e.g., H3K27me3) and early embryonic imprinting of the inactive X chromosome (Okamoto et al. 2004; Mak et al. 2004; Bao et al. 2005). Based on the above-mentioned considerations on the different stabilities of histone modifications, a repressive histone lysine trimethyl state appears as a more robust imprint that would ideally be suited to inherit long-term epigenetic information.

To address the different stability of epigenetic marks in the context of reprogramming potential, bovine embryos were analyzed after in vitro fertilization (IVF) and upon transfer of a fibroblast nucleus into an enucleated oocyte (Santos et al. 2003). Preimplantation embryos at the four-cell stage were then stained with antibodies against DNA methylation (5-me CpG) or with "multi-methyl" antibodies that detect repressive histone lysine trimethyl states. In normal embryos, only low levels of DNA methylation and no repressive histone methylation marks were detected (Fig. 5). In contrast, in cloned embryos, signals were enhanced for 5-me CpG and robust staining was observed with the multi-methyl histone antibodies, in a pattern that reflected the characteristic heterochromatic foci as they are present in the fibroblastic donor nuclei. These data indicate the persistence of soma-specific histone lysine trimethylation states in cloned embryos and argue that some repressive histone modifications may indeed be refractory to reprogramming by nuclear transfer.

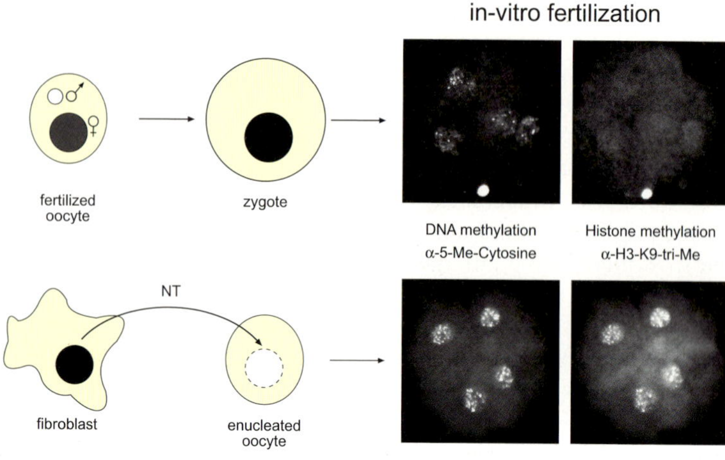

Fig. 5. Persistence of epigenetic modifications in cloned embryos. Epigenetic modifications were analyzed in bovine embryos generated either by in vitro fertilization (IVF; *top panel*) or by nuclear transfer (NT) of fibroblastic nuclei (*lower panel*). Preimplantation embryos at the four-cell stage were stained for DNA methylation with an antibody detecting 5-methyl CpG (5-me CpG) and for histone lysine methylation with a multi-methyl antibody that is specific for repressive trimethyl states. Modified after Santos et al. 2003

Most of the cloned embryos abort, suggesting that perturbed epigenetic imprints represent a major bottleneck for normal development and could be the cause for the poor efficiencies of ART (assisted reproductive technologies) and the reduced vigor of cloned animals (Rideout et al. 2001; Williams 2003). Failure to reprogram some of the more stable histone modifications would argue against the presence of trimethyl-specific histone lysine hydroxylases in early mammalian embryos, although these enzymes (if they exist) may be up-regulated upon implantation. In contrast, activating histone marks, such as acetylation or arginine methylation, appear highly dynamic during early mammalian development (Sarmento et al. 2004), consistent with elevated levels of HDACs and PADs in four-cell stage embryos (Wang et al. 2004).

1.6 Epigenetic Transitions During Tumorigenesis

The delicate balance between cell proliferation and normal differentiation is largely regulated by oncogenes transducing mitotic stimuli and by tumor suppressor genes surveying an intact genome function (Hanahan and Weinberg 2000). In addition, overall chromatin structure is controlled by several checkpoints that ensure DNA damage repair and prevent unequal chromosome segregation, thereby protecting genome integrity (Lengauer et al. 1998). Neoplastic transformation is almost invariably triggered by an apparently random pattern of activated (mutated) oncogenes and the silencing of tumor suppressor genes, resulting in the selective growth advantage of cancer cells. Loss of checkpoint controls would also allow survival of tumor cells despite accumulation of chromosomal aberrations and genomic aneuploidies. Although every cancer is primarily induced by genetic lesions (mutations, translocations, amplifications, etc.), there is a significant contribution of epigenetic modifications in altering the tumor spectrum or metastatic potential of transformed cells (Jones and Baylin 2002) and in preventing successful therapy by a variety of anti-cancer treatments (Marks et al. 2004). In addition, a large number of human diseases, ranging from proliferative defects to neurological disorders, have a strong epigenetic component (Jiang et al. 2004).

Epigenetic dysfunction in cancer progression has largely been attributed to perturbed DNA methylation, with broad hypomethylation (Feinberg et al. 2002; Bird 2002) along noncoding and repetitive sequences and localized hypermethylation at the promoters of many tumor suppressor genes (Jones and Baylin 2002). Whereas regional DNA hypomethylation would favor translocations and genomic instabilities, selective promoter DNA hypermethylation can result in the silencing of tumor suppressor genes. With the discoveries that histone modifications are functionally linked with DNA methylation (Fuks et al. 2000; Robertson et al. 2000; Rountree et al. 2000), and some histone marks, such as H3K9 methylation (Tamaru and Selker 2001; Jackson et al. 2002), may even direct DNA methylation, epigenetic transitions other than aberrant DNA methylation have gained significant interest for both diagnostic and therapeutic advances in cancer research. For example, reduced abundance of H4-K20me3 (Fraga et al. 2005) or perturbed histone acetylation profiles (Seligson et al. 2005) have been associated with neoplastic transformation, and many histone-modifying enzymes, such as HDACs, HMTases (Schneider et al. 2002) or Polycomb-group genes (Valk-Lingbeek et al. 2004) are mutated in human cancers.

Upon the transition of a normal cell to a cancer cell, a tumor suppressor gene, such as hMLH1 (Fahrner et al. 2002) or p16 (Nguyen

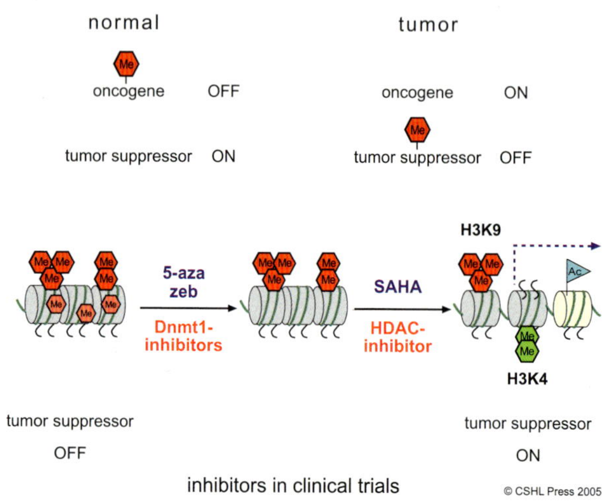

Fig. 6. Perturbed epigenetic control at tumor suppressor genes. *Top panel:* Neoplastic transformation is often associated with aberrant epigenetic control (e.g., DNA methylation and histone modifications) that results in inappropriate activation of oncogenes and abnormal silencing of tumor suppressor genes. *Bottom panel*: Treatment of tumor cells with the DNMT1 inhibitors 5-aza-cytidine (*5-aza*) or zebularine (*zeb*) leads to loss of DNA methylation at the promoter regions of silenced tumor suppressor genes without significantly altering histone lysine methylation patterns. Prolonged treatment with 5-aza, or a combination between DNMT1 and HDAC inhibitors (e.g., SAHA), facilitates loss of repressive H3K9me2, transcriptional reactivation and the induction of activating histone modifications (McGarvey et al., personal communication). Despite ongoing transcriptional activity, repressive H3K9me3 marks persist at the promoter and may trigger subsequent resilencing, thereby impairing efficient epigenetic therapy

et al. 2002), is silenced by both DNA methylation and repressive histone lysine methylation comprising H3K9me2, H3K9me3, and other marks. Inhibition of the major DNA methyltransferase (DNMT1) by 5-aza cytidine, or its analog zebularine (J.C. Cheng et al. 2004), removes DNA methylation but does not significantly alter H3K9 and H3K27 methylation patterns across the promoter region (Fig. 6). Only after prolonged 5-aza cytidine treatment or when combined with HDAC inhibitors (e.g., TSA or SAHA) will the promoter be reactivated, resulting in transcriptional stimulation and induction of activating marks, such as histone acetylation and H3K4 methylation. A major fraction of H3K9me2 and of repressive H3K27 methylation is lost, presumably by transcription-coupled histone variant exchange or by an LSD1-like demethylase activity, whereas, surprisingly, H3K9me3 marks persist (McGarvey et al., personal communication). Currently, it is unresolved whether the remaining H3K9me3 could induce subsequent re-silencing of the tumor suppressor gene and thereby interfere with durable epigenetic therapy, once the inhibitor treatment is paused. Despite these findings, zebularine (Yoo et al. 2004) and SAHA (Marks and Jiang 2005) are being used in phase I/II clinical trials, with good prognosis for several solid tumors and leukemias. Based on the complex interplay between DNA methylation and histone modifications, a combination therapy between DNMT1, HDAC and, probably, HMTase inhibitors could be an advantageous approach that may more efficiently reset aberrant epigenetic imprints in cancer therapy.

1.7 From Basic to Applied Research

In recent years, the many new insights and novel enzymatic activities have greatly advanced our understanding of the molecular mechanisms that govern epigenetic control. Are there common patterns that would be of predictive value and could be used for knowledge transfer from basic to applied research? Regarding the vast array of histone modifications and chromatin-modifying enzymes, it appears difficult, if not impossible, to assign a one-modification–one-disease link. However, for some of the more stable marks, in particular H3K9 and H4K20 trimethylation and for distinct combinations of histone modifications, there are specific staining profiles in interphase chromatin that can be correlated with diverse cell fates or with different proliferative potentials and that would be of diagnostic significance.

For a normal (differentiated) cell, H3K9me3 and H4K20me3 signals are detected in characteristic heterochromatic foci (around 15–20 in mouse interphase chromatin) that reflect pericentric regions of mammalian chromosomes (Peters et al. 2003; Schotta et al. 2004) (Fig. 7). Senescent or aged cells accumulate large clusters of "ectopic" heterochromatin (Narita et al. 2003) and have abnormal nuclear morphologies. Thus, elevated levels for H3K9me3 and H4K20me3 and reduced definition of histone acetylation reinforce repressive chromatin functions, block cellular plasticity, and can drive a cell into an antiproliferative state. In agreement with this model, *Suv39h* mutant cells, which lack pericentric H3K9me3, display higher immortalization rates while allowing genomic instabilities (Peters et al. 2001) and impair senescence

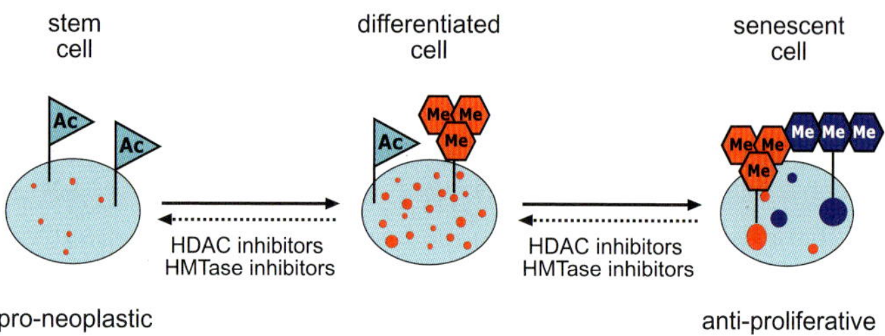

Fig. 7. Global patterns of histone modifications in diverse cell types. Normal, differentiated cells are characterized by a balanced representation of active (e.g., acetylation; *blue flag*) and repressive (e.g., lysine methylation; *red* and *blue hexagons*) histone modifications. In senescent or aged cells, repressive modifications accumulate at large blocks of ectopic heterochromatin. In contrast, stem cells display a general under-representation of repressive histone modifications that could be indicative of epigenetic plasticity. These global histone modification patterns can be used as diagnostic markers for cell fate specification. In addition, they suggest a possible combination therapy between HDAC and HMTase inhibitors to reprogram senescent cells or, on the other hand, to drive tumor cells into information overflow, chromatin catastrophe and apoptosis

in a *Ras*-driven tumor model (Braig et al. 2005). In contrast, increased levels of the Polycomb group HMTase Ezh2, which transduces H3K27 methylation, are correlated with a high risk for prostate cancer (Varambally et al. 2002).

Despite the pro-neoplastic state conferred by aberrant Ezh2 expression, there is a general reduction in global H3K9me3 and H4K20me3 (but a constitutively high level of histone acetylation) in cells that resembles an uncommitted or stem cell fate (Baxter et al. 2004). This correlation has recently also been shown for underrepresentation of H4K20me3 (Fraga et al. 2005) and for increased histone acetylation (Seligson et al. 2005) in human cancer cells, and is further reflected by screening tumor tissue banks for aberrant patterns of repressive histone lysine methylation states (K. Zatloukal and T. Jenuwein, unpublished data). Thus, low levels of H3K9me3 and H4K20me3, and the absence of the characteristic pericentric foci can be used as a diagnostic marker for neoplastic transformation in tumor progression. Dispersed nuclear distribution of HP1, which recognizes pericentric H3K9 methylation, has previously been described to correlate with the increased metastatic potential of tumor cells (Kirschmann et al. 2000).

There are around 50 predicted SET domain HMTases in the mammalian genome (Schneider et al. 2002; Schotta et al. 2004), and a significant proportion of the characterized enzymes, such as Suv39h, Ezh2, Mll, Nsd1, Riz and others, have been implicated in tumor development (Schneider et al. 2002). As enzymes that require a substrate (histone peptide) and a co-factor (S-adenosyl methionine, SAM), they are ideally suited as novel targets in high-throughput screens (HTSs) to develop small molecule inhibitors which could then be used in cancer therapy or exploratory research to convert a committed cell to a more undifferentiated or stem cell character. Similar approaches have been useful for HDAC and DNMT1 inhibitors, and HTSs were recently performed to uncover compounds that modulate PRMTs (D. Cheng et al. 2004) or identified a natural fungal substance, chaetocin, which appears to reduce HMTase activity for H3K9 dimethylation (Greiner et al. 2005). For HMTases, there are the extra complications that histone lysine methylation can be either activating (e.g., H3K4) or repressing (e.g., H3K9), and the respective enzymes vary in their potential to induce mono-, di- or trimethylation and may, in addition, even have non-histone substrates

(Chuikov et al. 2004). It remains to be seen whether some of the other HTSs (e.g., S. Kubicek, T. Jenuwein, and Boehringer Ingelheim, unpublished data) will be selective enough to derive inhibitors that can discriminate between PRMTs and HMTases and also differ in their efficacy to block histone lysine mono-, di- or trimethylation. If these specific requirements can be met, then HMTase inhibitors, or combinations between HDAC and HMTase inhibitors, will represent an attractive novel avenue for cancer therapy and the potential use in tissue regeneration and stem cell research.

For example, a selective inhibitor against MLL (Okada et al. 2005) or EZH2 (Varambally et al. 2002) would be a valuable agent in the projected epigenetic therapy for leukemia or prostate cancer, while it should not interfere with the activity of the SUV39H HMTases. The use of SUV39H inhibitors initially appears to be counter-intuitive because they may lead to chromosome instabilities (Peters et al. 2001) or otherwise impair the tumor suppressor function of the Suv39h enzymes (Braig et al. 2005). However, they may sensitize tumor cells more than normal cells, as is also observed for the generic microtubule destabilizing agent taxol (Zhou and Giannakakou 2005) or the anti-cancer drug cisplatin (Wang and Lippard 2005). For the knowledge transfer between basic and applied research, both hypothesis-driven and empirical approaches are required to ultimately define the efficacy and usefulness of the potential HMTase inhibitors. It is based on these considerations that a combination of an HDAC and an HMTase inhibitor may be more potent than a single drug treatment to reprogram a senescent or differentiated cell to an uncommitted and plastic state (Fig. 7). On the other hand, such a combination therapy could also drive a pro-neoplastic cell into information overflow and chromatin catastrophe, thereby killing more tumor cells than normal cells. For example, we also do not know whether some of the clinically relevant HDAC inhibitors primarily work through transcriptional reactivation of certain target genes or function by sensitizing chromatin lesions, inhibit efficient DNA repair and potentiate genomic instabilities, which then lead to apoptosis of tumor cells.

In summary, the many mechanistic advances in epigenetic control over the recent years have not only extended our understanding of the broad implications of chromatin regulation but, more specifically, also offer great potential for novel diagnostic and therapeutic avenues in addressing some of the long-standing questions, such as cell type differentiation, stem cell plasticity, tumorigenesis, and even aging.

Acknowledgements. We would like to thank David Allis, Robin Allshire, Steve Baylin, Rudolf Jaenisch, Peter Jones, Tony Kouzarides, Gunter Reuter, and Kurt Zatloukal for insightful discussions and for allowing us to cite work prior to its publication. Research in the laboratory of T. Jenuwein is supported by the IMP through Boehringer Ingelheim and by grants from the European Union (FP6 NoE-network "The Epigenome") and the Austrian GEN-AU initiative which is financed by the Austrian Ministry of Education, Science and Culture.

References

Arney KL, Bao S, Bannister AJ, Kouzarides T, Surani MA (2002) Histone methylation defines epigenetic asymmetry in the mouse zygote. Int J Dev Biol 46:317–320

Bao S, Miyoshi N, Okamoto I, Jenuwein T, Heard E, Surani MA (2005) Initiation of epigenetic reprogramming of the X chromosome in somatic nuclei transplanted to a mouse oocyte. EMBO Rep 6:748–754

Baxter J, Sauer S, Peters A, John R, Williams R, Caparros ML, Arney K, Otte A, Jenuwein T, Merkenschlager M, Fisher AG (2004) Histone hypomethylation is an indicator of epigenetic plasticity in quiescent lymphocytes. EMBO J 23:4462–4472

Berger SL (2002) Histone modifications in transcriptional regulation. Curr Opin Genet Dev 12:142–148

Bird A (2002) DNA methylation patterns and epigenetic memory. Genes Dev 16:6–21

Braig M, Lee S, Loddenkemper C, Rudolph C, Peters AH, Stein H, Dorken B, Schlegelberger B, Jenuwein T, Schmitt CA (2005) Oncogene-induced senescence as an initial barrier in lymphoma development. Nature 436:660–665

Briggs SD, Xiao T, Sun ZW, Caldwell JA, Shabanowitz J, Hunt DF, Allis CD, Strahl BD (2002) Gene silencing: trans-histone regulatory pathway in chromatin. Nature 418:498

Cavalli G, Paro R (1999) Epigenetic inheritance of active chromatin after removal of the main transactivator. Science 286:955–958

Cheng JC, Yoo CB, Weisenberger DJ, Chuang J, Wozniak C, Liang G, Marquez VE, Greer S, Orntoft TF, Thykjaer T, Jones PA (2004) Preferential response of cancer cells to zebularine. Cancer Cell 6:151–158

Cheng D, Yadav N, King RW, Swanson MS, Weinstein EJ, Bedford MT (2004) Small molecule regulators of protein arginine methyltransferases. J Biol Chem 279:23892–23899

Chuikov S, Kurash JK, Wilson JR, Xiao B, Justin N, Ivanov GS, McKinney K, Tempst P, Prives C, Gamblin SJ, Barlev NA, Reinberg D (2004) Regulation of p53 activity through lysine methylation. Nature 432:353–360

Cuthbert GL, Daujat S, Snowden AW, Erdjument-Bromage H, Hagiwara T, Yamada M, Schneider R, Gregory PD, Tempst P, Bannister AJ, Kouzarides T (2004) Histone deimination antagonizes arginine methylation. Cell 118:545–553

Daniel JA, Pray-Grant MG, Grant PA (2005) Effector proteins for methylated histones: an expanding family. Cell Cycle 4:919–926

Dhalluin C, Carlson JE, Zeng L, He C, Aggarwal AK, Zhou M-M (1999) Structure and ligand of a histone acetyltransferase bromodomain. Nature 399:491–496

Fahrner JA, Eguchi S, Herman JG, Baylin SB (2002) Dependence of histone modifications and gene expression on DNA hypermethylation in cancer. Cancer Res 62:7213–7238

Feinberg AP, Cui H, Ohlsson R (2002) DNA methylation and genomic imprinting: insights from cancer into epigenetic mechanisms. Semin Cancer Biol 12:389–398

Fischle W, Wang Y, Allis CD (2003) Binary switches and modification cassettes in histone biology and beyond. Nature 425:475–479

Fraga MF, Ballestar E, Villar-Garea A, Boix-Chornet M, Espada J, Schotta G, Bonaldi T, Haydon C, Ropero S, Petrie K, Iyer NG, Perez-Rosado A, Calvo E, Lopez JA, Cano A, Calasanz MJ, Colomer D, Piris MA, Ahn N, Imhof A, Caldas C, Jenuwein T, Esteller M (2005) Loss of acetylation at Lys16 and trimethylation at Lys20 of histone H4 is a common hallmark of human cancer. Nat Genet 37:391–400

Freitas MA, Sklenar AR, Parthun MR (2003) Application of mass spectrometry to the identification and quantification of histone post-translational modifications. J Cell Biochem 92:691–700

Fuks F, Burgers WA, Brehm A, Hughes-Davies L, Kouzarides T (2000) DNA methyltransferase Dnmt1 associates with histone deacetylase activity. Nat Genet 24:88–91

Goffeau A, Barrell BG, Bussey H, Davis RW, Dujon B, Feldmann H, Galibert F, Hoheisel JD, Jacq C, Johnston M, Louis EJ, Mewes HW, Murakami Y, Philippsen P, Tettelin H, Oliver SG (1996) Life with 6000 genes. Science 274:546–567

Greiner D, Bonaldi T, Eskeland R, Roemer E, Imhof A (2005) Identification of a specific inhibitor of the histone methyltransferase SU(VAR)3-9. Nat Chem Biol 1:143–145

Grewal SIS, Klar AJS (1996) Chromosomal inheritance of epigenetic states in fission yeast during mitosis and meiosis. Cell 86:95–101

Hanahan D, Weinberg RA (2000) The hallmarks of cancer. Cell 100:57–70
Henikoff S, Ahmad K (2005) Assembly of variant histones into chromatin. Annu Rev Cell Dev Biol advance online publication, 18 July 2005 (doi:10.1146/annurev.cellbio.21.012704.133518)
Hochedlinger K, Blelloch R, Brennan C, Yamada Y, Kim M, Chin L, Jaenisch R (2004) Reprogramming of a melanoma genome by nuclear transplantation. Genes Dev 18:1875–1885
Jackson JP, Lindroth AM, Cao X, Jacobsen SE (2002) Control of CpNpG DNA methylation by the KRYPTONITE histone H3 methyltransferase. Nature 416:556–560
Jacobson RH, Ladurner AG, King DS, Tjian R (2000) Structure and function of a human TAFII250 double bromodomain module. Science 288:1422–1425
Jaenisch R, Bird A (2003) Epigenetic regulation of gene expression: how the genome integrates intrinsic and environmental signals. Nat Genet 33:245–254
Jaenisch R, Hochedlinger K, Eggan K (2005) Nuclear cloning, epigenetic reprogramming and cellular differentiation. Novartis Found Symp 265:107–118
Jenuwein T, Allis CD (2001) Translating the histone code. Science 293:1074–1080
Jiang Y-h, Bressler J, Beaudet AL (2004) Epigenetics and human disease. Annu Rev Genomics Hum Genet 5:479–510
Jones PA, Baylin SB (2002) The fundamental role of epigenetic events in cancer. Nat Rev Genet 3:415–428
Kirschmann DA, Lininger RA, Gardner LM, Seftor EA, Odero VA, Ainsztein AM, Earnshaw WC, Wallrath LL, Hendrix MJ (2000) Down-regulation of $HP1^{Hs\alpha}$ expression Is associated with the metastatic phenotype in breast cancer. Cancer Res 60:3359–3363
Klochendler-Yeivin A, Muchardt C, Yaniv M (2002) SWI/SNF chromatin remodeling and cancer. Curr Opin Genet Dev 12:73–79
Kuzmichev A, Jenuwein T, Tempst P, Reinberg D (2004) Different Ezh2-containing complexes target methylation of histone H1 or nucleosomal histone H3. Mol Cell 14:183–193
Lachner M, O'Sullivan RJ, Jenuwein T (2003) An epigenetic road map for histone lysine methylation. J Cell Sci 116:2117–2124
Lander ES, Linton LM, Birren B et al. (International Human Genome Sequencing Consortium) (2001) Initial sequencing and analysis of the human genome. Nature 409:860–921
Lengauer C, Kinzler KW, Vogelstein B (1998) Genetic instabilities in human cancers. Nature 396:643–649
Lippman Z, Martienssen R (2004) The role of RNA interference in heterochromatic silencing. Nature 431:364–370

Luger K, Mader AW, Richmond RK, Sargent DF, Richmond TJ (1997) Crystal structure of the nucleosome core particle at 2.8 Å resolution. Nature 389:251–260

Mak W, Nesterova TB, de Napoles M, Appanah R, Yamanaka S, Otte AP, Brockdorff N (2004) Reactivation of the paternal X chromosome in early mouse embryos. Science 303:666–669

Marks PA, Jiang X (2005) Histone deacetylase inhibitors in programmed cell death and cancer therapy. Cell Cycle 4:549–551

Marks PA, Richon VM, Miller T, Kelly WK (2004) Histone deacetylase inhibitors. Adv Cancer Res 91:137–168

Maurer-Stroh S, Dickens NJ, Hughes-Davies L, Kouzarides T, Eisenhaber F, Ponting CP (2003) The tudor domain 'royal family': tudor, plant agenet, chromo, PWWP and MBT domains. Trends Biochem Sci 28:69–74

Metzger E, Wissmann M, Yin N, Muller JM, Schneider R, Peters AH, Gunther T, Buettner R, Schule R (2005) LSD1 demethylates repressive histone marks to promote androgen-receptor-dependent transcription. Nature advance online publication, 3 August 2005 (doi:10.1038/nature04020)

Morgan HD, Santos F, Green K, Dean W, Reik W (2005) Epigenetic reprogramming in mammals. Hum Mol Genet 14:R47–R58

Narita M, Nunez S, Heard E, Narita M, Lin AW, Hearn SA, Spector DL, Hannon GJ, Lowe SW (2003) Rb-mediated heterochromatin formation and silencing of E2F target genes during cellular senescence. Cell 113:703–716

Narlikar GJ, Fan H-Y, Kingston RE (2002) Cooperation between complexes that regulate chromatin structure and transcription. Cell 108:475–487

Ng HH, Feng Q, Wang H, Erdjument-Bromage H, Tempst P, Zhang Y, Struhl K (2002) Lysine methylation within the globular domain of histone H3 by Dot1 is important for telomeric silencing and Sir protein association. Genes Dev 16:1518–1527

Nguyen CT, Weisenberger DJ, Velicescu M, Gonzales FA, Lin JC, Liang G, Jones PA (2002) Histone H3-Lysine 9 methylation is associated with aberrant gene silencing in cancer cells and is rapidly reversed by 5-Aza-2′-deoxycytidine. Cancer Res 62:6456–6461

Nishioka K, Rice JC, Sarma K, Erdjument-Bromage H, Werner J, Wang Y, Chuikov S, Valenzuela P, Tempst P, Steward R, Lis JT, Allis CD, Reinberg D (2002) PR-SET7 is a nucleosome-specific methyltransferase that modifies lysine 20 of histone H4 and is associated with silent chromatin. Mol Cell 9:1201–1213

Okada Y, Feng Q, Lin Y, Jiang Q, Li Y, Coffield VM, Su L, Xu G, Zhang Y (2005) hDOT1L links histone methylation to leukemogenesis. Cell 121:167–178

Okamoto I, Otte AP, Allis CD, Reinberg D, Heard E (2004) Epigenetic dynamics of imprinted X inactivation during early mouse development. Science 303:644–649

Peters AH, O'Carroll D, Scherthan H, Mechtler K, Sauer S, Schofer C, Weipoltshammer K, Pagani M, Lachner M, Kohlmaier A, Opravil S, Doyle M, Sibilia M, Jenuwein T (2001) Loss of the Suv39h histone methyltransferases impairs mammalian heterochromatin and genome stability. Cell 107:323–337

Peters AH, Kubicek S, Mechtler K, O'Sullivan RJ, Derijck AA, Perez-Burgos L, Kohlmaier A, Opravil S, Tachibana M, Shinkai Y, Martens JH, Jenuwein T (2003) Partitioning and plasticity of repressive histone methylation states in mammalian chromatin. Mol Cell 12:1577–1589

Rice JC, Briggs SD, Ueberheide B, Barber CM, Shabanowitz J, Hunt DF, Shinkai Y, Allis CD (2003) Histone methyltransferases direct different degrees of methylation to define distinct chromatin domains. Mol Cell 12:1591–1598

Rideout WM III, Eggan K, Jaenisch R (2001) Nuclear cloning and epigenetic reprogramming of the genome. Science 293:1093–1098

Ringrose L, Paro R (2004) Epigenetic regulation of cellular memory by the polycomb and trithorax group proteins. Annu Rev Genet 38:413–443

Robertson KD, Ait-Si-Ali S, Yokochi T, Wade PA, Jones PL, Wolffe AP (2000) DNMT1 forms a complex with Rb, E2F1 and HDAC1 and represses transcription from E2F-responsive promoters. Nat Genet 25:338–342

Rountree MR, Bachman KE, Baylin SB (2000) DNMT1 binds HDAC2 and a new co-repressor, DMAP1, to form a complex at replication foci. Nat Genet 25:269–277

Sanders SL, Portoso M, Mata J, Bahler J, Allshire RC, Kouzarides T (2004) Methylation of histone H4 lysine 20 controls recruitment of Crb2 to sites of DNA damage. Cell 119:603–614

Santos F, Peters AH, Otte AP, Reik W, Dean W (2005) Dynamic chromatin modifications characterize the first cell cycle in mouse embryos. Dev Biol 280:225–236

Santos F, Zakhartchenko V, Stojkovic M, Peters A, Jenuwein T, Wolf E, Reik W, Dean W (2003) Epigenetic marking correlates with developmental potential in cloned bovine preimplantation embryos. Curr Biol 13:1116–1121

Sarma K, Reinberg D (2005) Histone variants meet their match. Nat Rev Mol Cell Biol 6:139–149

Sarmento OF, Digilio LC, Wang Y, Perlin J, Herr JC, Allis CD, Coonrod SA (2004) Dynamic alterations of specific histone modifications during early murine development. J Cell Sci 117:4449–4459

Schneider R, Bannister AJ, Kouzarides T (2002) Unsafe SETs: histone lysine methyltransferases and cancer. Trends Biochem Sci 27:396–402

Schotta G, Lachner M, Sarma K, Ebert A, Sengupta R, Reuter G, Reinberg D, Jenuwein T (2004) A silencing pathway to induce H3-K9 and H4-K20 trimethylation at constitutive heterochromatin. Genes Dev 18:1251–1262

Schreiber SL, Bernstein BE (2002) Signaling network model of chromatin. Cell 111:771–778

Seligson DB, Horvath S, Shi T, Yu H, Tze S, Grunstein M, Kurdistani SK (2005) Global histone modification patterns predict risk of prostate cancer recurrence. Nature 435:1262–1266

Shi Y, Lan F, Matson C, Mulligan P, Whetstine JR, Cole PA, Casero RA, Shi Y (2004) Histone demethylation mediated by the nuclear amine oxidase homolog LSD1. Cell 119:941–953

Strahl BD, Allis CD (2000) The language of covalent histone modifications. Nature 403:41–45

Tamaru H, Selker EU (2001) A histone H3 methyltransferase controls DNA methylation in Neurospora crassa. Nature 414:277–283

Trewick SC, McLaughlin PJ, Allshire RC (2005) Methylation: lost in hydroxylation? EMBO Rep 6:315–320

Turner BM (2000) Histone acetylation and an epigenetic code. BioEssays 22:836–845

Valk-Lingbeek ME, Bruggeman SW, van Lohuizen M (2004) Stem cells and cancer: the Polycomb connection. Cell 118:409–418

Van der Heijden GW, Dieker JW, Derijck AA, Muller S, Berden JH, Braat DD, van der Vlag J, de Boer P (2005) Asymmetry in histone H3 variants and lysine methylation between paternal and maternal chromatin of the early mouse zygote. Mech Dev 122:1008–1022

Van Holde KE (1988) Chromatin. Springer, Berlin Heidelberg New York

Van Leeuwen F, Gafken PR, Gottschling DE (2002) Dot1p modulates silencing in yeast by methylation of the nucleosome core. Cell 109:745–756

Varambally S, Dhanasekaran SM, Zhou M, Barrette TR, Kumar-Sinha C, Sanda MG, Ghosh D, Pienta KJ, Sewalt RG, Otte AP, Rubin MA, Chinnaiyan AM (2002) The polycomb group protein EZH2 is involved in progression of prostate cancer. Nature 419:624–629

Wang D, Lippard SJ (2005) Cellular processing of platinum anticancer drugs. Nat Rev Drug Discov 4:307–320

Wang Y, Wysocka J, Sayegh J, Lee YH, Perlin JR, Leonelli L, Sonbuchner LS, McDonald CH, Cook RG, Dou Y, Roeder RG, Clarke S, Stallcup MR, Allis CD, Coonrod SA (2004) Human PAD4 regulates histone arginine methylation levels via demethylimination. Science 306:279–283

Waterston RH et al. (Mouse Genome Sequencing Consortium) (2002) Initial sequencing and comparative analysis of the mouse genome. Nature 420:520–562

Williams N (2003) Death of Dolly marks cloning milestone. Curr Biol 13:R209–R10

Wysocka J, Swigut T, Milne TA, Dou Y, Zhang X, Burlingame AL, Roeder RG, Brivanlou AH, Allis CD (2005) WDR5 associates with histone H3 methylated at K4 and is essential for H3 K4 methylation and vertebrate development. Cell 121:859–872

Xiao B, Jing C, Kelly G, Walker PA, Muskett FW, Frenkiel TA, Martin SR, Sarma K, Reinberg D, Gamblin SJ, Wilson JR (2005) Specificity and mechanism of the histone methyltransferase Pr-Set7. Genes Dev 19:1444–1454

Xu F, Zhang K, Grunstein M (2005) Acetylation in histone H3 globular domain regulates gene expression in yeast. Cell 121:375–385

Ye J, Ai X, Eugeni EE, Zhang L, Carpenter LR, Jelinek MA, Freitas MA, Parthun MR (2005) Histone H4 lysine 91 acetylation: a core domain modification associated with chromatin assembly. Mol Cell 18:123–130

Yoo CB, Cheng JC, Jones PA (2004) Zebularine: a new drug for epigenetic therapy. Biochem Soc Trans 32:910–912

Zhang X, Yang Z, Khan SI, Horton JR, Tamaru H, Selker EU, Cheng X (2003) Structural basis for the product specificity of histone lysine methyltransferases. Mol Cell 12:177–185

Zhou J, Giannakakou P (2005) Targeting microtubules for cancer chemotherapy. Curr Med Chem Anti-Canc Agents 5:65–71

2 Nucleosome Structure and Function

J.V. Chodaparambil, R.S. Edayathumangalam, Y. Bao,
Y.-J. Park, K. Luger

Abstract. It is now widely recognized that the packaging of genomic DNA, together with core histones, linker histones, and other functional proteins into chromatin profoundly influences nuclear processes such as transcription, replication, DNA repair, and recombination. How chromatin structure modulates the expression of knowledge encoded in eukaryotic genomes, and how these processes take place within the context of a highly complex and compacted genomic chromatin environment remains a major unresolved question in biology. Here we review recent advances in nucleosome structure and dynamics.

2.1 Introduction

Virtually all in vitro studies of processes that require access to eukaryotic DNA have to take into account its packaging with a roughly equal mass of proteins to form large macromolecular assemblages termed chromatin. Histones, by forming nucleosomes, are responsible for the first level of structural organization (Luger et al. 1997a). Hundreds of thousands of nucleosomes are further compacted into hierarchical structures of increasing complexity and largely unknown architecture (Luger and Hansen 2005). Nonhistone proteins and ions are significantly involved in the formation of these higher-order structures. Chromatin and nucleosomes are inherently dynamic and highly malleable, allowing them to accomplish two conflicting yet vital tasks: first, the approximately 2 m of eukaryotic DNA have to be packaged within the confines of the nucleus, while preventing knots and tangles. Second, the information that is encoded within the DNA needs to be accessed at appropriate times, and this is to a large extent regulated by local changes in chromatin structure. Histones have evolved to accommodate these two conflicting functions to perfection. The nucleosome is designed to highly constrain DNA to achieve approximately fivefold compaction at the first level and to promote the formation of higher-order structure; yet several independent mechanisms ensure chromatin fluidity to allow highly regulated access to the packaged DNA.

2.2 Nucleosome Structure

The nucleosome is the elemental repeating unit in chromatin, consisting of a protein core (the histone octamer) around which 146 base pairs of DNA are wrapped in nearly two turns of a tight superhelix (Luger et al. 1997a). The protein core itself is composed of two copies each of the four core histone proteins H2A, H2B, H3, and H4.

2.2.1 Histone Fold

The core histones share a structurally conserved motif, the histone fold, averaging about 70 amino acids (Arents and Moudrianakis 1995). The histone fold consists of three α-helices connected by short loops. Two short, 10–14 residue helices flank a longer, 28-amino-acid-long central α helix, creating a shallow slot. In solution, H2A always forms a tight heterodimer with H2B, and H3 with H4 (Fig. 1A,B, and C,D, respectively). These heterodimers are stabilized through large, mostly hydrophobic interaction interfaces formed by the antiparallel arrangement of the two long α2 helices. Specificity of dimerization is given by the complementarity of amino acids in both subunits. Dimerization leads to the alignment of the L1 loop of one histone with the L2 loop of the other at the edges of the crescent-shaped dimer, and a close juxtaposition of the N-termini of the two α1 helices at its apex. This arrangement forms three DNA interaction modules per histone fold dimer, two L1L2 sites, and one α1α1 binding site (Luger et al. 1997a). Despite the low degree of sequence homology between the amino acid sequences of the four core histone proteins, their histone fold motifs can be superimposed with little structural divergence (Fig. 1A and C). The histone fold motif has been found in a variety of transcription factors and in subunits of large macromolecular assemblies such as histone modification and chromatin remodeling complexes; however, none of this DNA binding has been demonstrated conclusively (Luger and Richmond 1998a).

2.2.2 Histone Fold Extensions

Histone fold extensions are structured regions outside the histone fold that are mainly responsible for protein–protein interactions within the histone octamer and for defining the surface of the nucleosome core particle. They also contribute to DNA binding. These extensions are structurally not conserved between the four core histones (although they are just as highly conserved between species as the histone fold regions), and are variable in length (Fig. 1B,D). They are involved in forming the interaction interface between H2A-H2B dimer and $(H3\text{-}H4)_2$ tetramer (for example, the H2A docking domain, or the C-terminal β-strand of H4), contribute to the organization of the DNA (e.g., the αN helix of H3 and of H2A), or help define the nucleosomal surface (H2B αC; Fig. 1B).

The historic functional partition of histones into "globular domains" and "histone tails" is derived from experiments in which nucleosomes were treated with limiting amounts of trypsin (Whitlock and Stein 1978). Whatever was removed from the histones through limited proteolysis

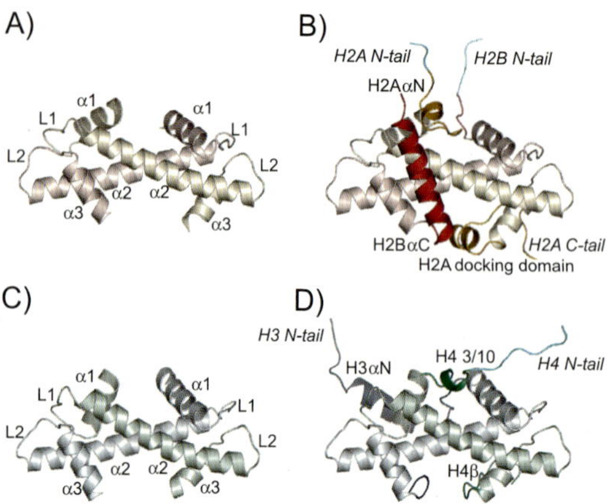

Fig. 1A–D. The overall structure of individual histone fold pairs is remarkably similar. **A** H2A-H2B, histone fold region only (shown in *light yellow* and *light red*, respectively). **B** H2A-H2B including histone fold extensions (shown in *yellow* and *red*) and tails (shown in *cyan*). **C** H3-H4 histone fold dimer, histone fold region only (*light blue* and *light green*, respectively). **D** H3-H4 including histone fold extensions and tails, shown in *dark blue* and *dark green*, respectively; (tails in *cyan*)

was termed a tail and was subsequently found to be dispensable for in vitro formation of nucleosomes, whereas the globular domains were responsible for maintaining the structure and stability of the nucleosome (Van Holde 1988). In light of several high-resolution nucleosome structures (reviewed in Luger 2003), we now know that the globular domains are definitively *not* globular, nor does the term "histone fold" encompass the entire structured region of each histone (Fig. 1A–D). We can also more sensibly define the histone tails as those parts of the histone proteins that extend from the confines of the DNA superhelix (Luger and Richmond 1998b).

2.2.3 Histone Tails

In all structures published to date, the histone tails are mostly disordered, with the exception of one H4 tail that engages in crystal structures. They do not form an integral part of the histone octamer, but may contribute minimally to DNA binding in the context of a mono-nucleosome. The histone tails are also the regions of highest sequence divergence among species, especially for H2A and H2B (Sullivan et al. 2002). Their true function is unlikely to be evident in the limited context of a mononucleosome, since they are thought to be involved in the formation of higher order structure (Dorigo et al. 2003). The more definitive structural delineation coincides quite well with the earlier definition derived from proteolysis, with the notable exception of histone H3 tail (earlier delineations: 1–23, structural delineation: 1–37; due to the absence of protease cleavage sites in the region between 23 and 37, Fig. 1B). Nucleosomes reconstituted with trypsinized or partially trypsinized core histones, or with the earlier version of recombinant H3 (Luger et al. 1997b) were historically used to investigate the role of histone tails in a variety of in vitro assays. When interpreting these data, it must be kept in mind that these nucleosomes still retained a sizable piece of the H3 tail (including several potential modification sites).

Covalent modification of the core histones (and variations in the fundamental biochemical composition of nucleosomes by the incorporation of histone variants) distinguishes transcriptionally active from inactive chromatin regions, by acting through mechanisms involving changing the structure of the nucleosomes, altering their ability to interact with other protein factors, modifying their propensity to fold into varying degrees of higher order structures, or by any combination of the above (Luger and Richmond 1998b). While the biological importance of histone post-translational modifications has been unequivocally established over the last decade, it is not at all clear from a structural standpoint how relatively minor modifications that arise from the post-translational addition of acetyl-, methyl-, or phosphate groups to the flexible histone tails can cause the observed dramatic biological effects. The earlier view that acetylation or phosphorylation of lysine residues changes the affinity of histone tails for the DNA and thus destabilizes the nucleosome, resulting in increased transcription rates, is definitively too simplistic.

Three fundamentally different, but not mutually exclusive roles for histone tails and their modifications have emerged. First, at the level of the mononucleosome, the histone tails are thought to modulate accessibility of nucleosomal DNA for base-specific recognition. Second, histone tails serve as interaction platforms for cellular factors (reviewed in Jenuwein and Allis 2001; Berger 2002; Strahl and Allis 2000). Third, they are to a large extent responsible for the formation of chromatin higher-order structure(s) by forming protein–protein interaction with neighboring nucleosomes to foster condensation of chromatin fibers (Dorigo et al. 2003). All of these functions can potentially be affected by post-translational modifications of amino acid side chains within the histone tails, by changing histone tail interactions with their respective binding partners.

2.2.4 DNA Structure

The massive distortion of DNA into a superhelix is brought about by the tight interaction between the rigid framework of the histone octamer with the DNA at 14 independent DNA binding sites (Luger and Richmond 1998a; numbered 1–7 in Fig. 2 depicting one-half of the nucleosome structure). The flat, wedge-like surface of the histone octamer displays a continuous basic groove along the surface around which the DNA is wound. Interactions between protein and DNA are made exclusively at the minor groove of the DNA, and are characterized by extensive hydrogen bonding between the protein's main chain amide atoms and the phosphate oxygen atoms of the DNA backbone. No base-specific contacts are observed. Water molecules contribute significantly to the interaction between protein and DNA (Davey et al. 2002). Structural analyses also hinted for the first time at a property of nucleosomal DNA that is now viewed as an essential characteristic: the dynamic or static displacement of DNA phosphates (as expressed in the crystallographic Debye–Waller temperature factors) is much higher than expected in regions where the DNA is solvent exposed, but much lower in regions where the phosphates interact with histones, following an approximately ten-base pair periodicity (Luger et al. 1997a). This and other data indicate that despite being highly confined by periodic interactions of both DNA strands with the histone octamer, nucleosomal DNA possesses a surprisingly high degree of freedom to accommodate ligands or to better adjust to positioning signals (Suto et al. 2003; Edayathumangalam et al. 2005).

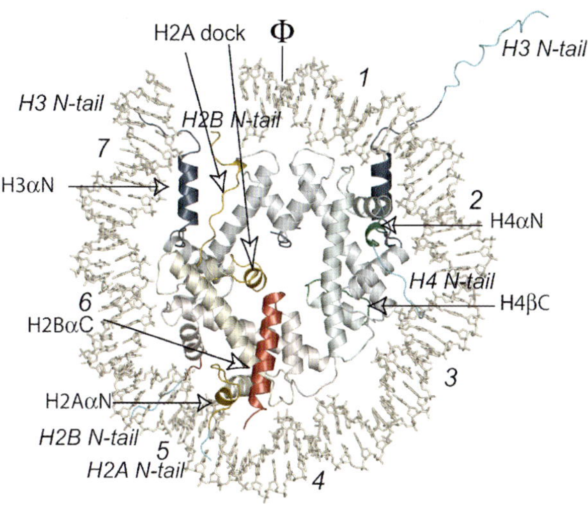

Fig. 2. Overview of the nucleosome structure. Only one-half of the DNA is shown (viewed down the superhelical axis, with associated proteins. Coloring as in Fig. 1. Italic numbers indicate regions of contact between the DNA minor groove and the histone octamer

Nucleosomal DNA is highly distorted and partially occluded from the solvent due to its tight interaction with the histone octamer (Fig. 2). Thus, nucleosome architecture greatly affects the accessibility of nucleosomal DNA for global and specific protein/macromolecular regulators.

2.2.5 Higher-Order Structures

Although we now have very detailed knowledge of the structure of nucleosomes from a variety of species and containing mutants and histone variants (reviewed in Luger 2003; Tsunaka et al. 2005), molecular details of the multiple levels of chromatin higher-order structure have remained elusive; nor do we know how global and specific regulators (for example, linker histone H1, or transcription factors) access DNA in the various structural contexts. These two questions represent two of the most pressing and difficult challenges in the structural biology of the nucleus. Testable models of chromatin higher-order structure and of nucleosomes in complex with global and specific regulators are essential for our understanding of how essential processes such as replication, transcription, repair, and recombination function in vivo.

Linker histones and other nonhistone proteins promote or stabilize the folding of nucleosomal arrays into superstructures of increasing complexity and largely unknown architecture (Hansen 2002). Recent inroads have been made into this challenging problem (Dorigo et al. 2004). A steadily growing number of activities have been identified that have the potential to alter chromatin structure; however, it is still for the most part unknown which structures are being interconverted, and by which mechanisms.

2.3 Nucleosome Dynamics

Our view of nucleosomes as monolithic and prohibitive structures in the path of DNA and RNA polymerases (Kornberg and Lorch 1991) is rapidly changing with the growing realization that nucleosomal DNA is highly flexible and adaptable (see, for example, Polach and Widom 1995; Suto et al. 2003; White and Luger 2004; Li and Widom 2004; Li et al. 2005) (Fig. 3A) and that histone subunits are exchanged at a rapid rate even in the absence of transcription and replication (Jackson 1990; Kimura and Cook 2001; Park et al. 2004) (Fig. 3B). The demonstration that the histone octamer is capable of sliding along the DNA over significant distances has further changed our perception of chromatin (reviewed in Flaus and Owen-Hughes 2003a, 2004). This knowledge is in apparent conflict with the known nucleosome architecture, which shows that approximately 120 direct (and an equal number of indirect) contacts exist between the histone octamer and the DNA over its entire length (Richmond and Davey 2003). Nucleosome sliding occurs in vivo, where the energy barrier toward nucleosome movement is overcome by ATP-dependent chromatin remodeling factors and/or by histone "chaperones" that transiently remove histone subunits from the nucleosome. It likely affects the accessibility and recognition of nucleosomal DNA by sequence-specific transcription factors.

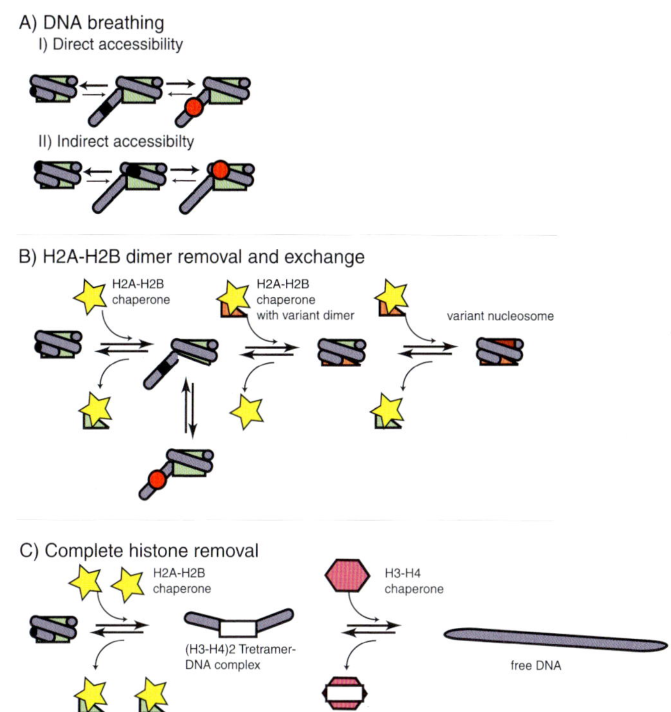

Fig. 3A–C. Mechanisms for modulating accessibility of nucleosomal DNA. Note that ATP-dependent chromatin remodeling factors are not discussed here. **A** DNA breathing according to the site-exposure model proposed by Widom and colleagues (Polach and Widom 1996; Li et al. 2004). **B** Chaperone-mediated histone removal and exchange. Dimers containing H2A variants are shown as *red triangles*, H2A-H2B dimer-specific chaperones are shown as *yellow stars*. **C** Complete histone removal by the sequential action of H2A-H2B and (H3-H4)-specific histone chaperones (*yellow stars* and *pink octagons*, respectively). Note that H2A-H2B dimers are also removed by passing RNA polymerase and by FACT (Kireeva et al. 2002; Belotserkovskaya et al. 2003)

2.3.1 DNA "Breathing"

In our laboratory, we have used several independent approaches to demonstrate the dynamic behavior of nucleosomal DNA in vitro. In the first example, we have used a combination of X-ray crystallography, DNase and hydroxyl radical footprinting, and affinity cleavage to show that in solution, nucleosomal DNA appears to dynamically alter its position with respect to the histone octamer in order to optimize the location of the positioning signals embedded in the sequence with respect to the histone octamer (Edayathumangalam et al. 2005). This dynamic behavior occurs independently of ATP-dependent chromatin remodeling factors or histone chaperones. The DNA sequence used for these experiments is a palindromic fragment derived from human α-satellite DNA (Luger et al. 1997a). This DNA sequence was originally chosen for crystallographic studies due to its outstanding positioning properties (that is, the DNA is positioned with high precision both translationally and rotationally). It is thus likely that other sequences, in which positioning signals are weaker or more contradictory, will exhibit an even more pronounced propensity to dynamically reposition themselves with respect to the histone octamer. The inherent conformational flexibility of nucleosomal DNA may assist the function of ATP-dependent chromatin remodeling factors and access of cellular regulatory factors.

We have also studied the structural consequences of DNA recognition and binding in the context of a defined mononucleosome. We observed that the ends of the DNA are capable of partial and reversible "unpeeling" from the surface of the histone octamer upon binding of a ligand or protein near the nucleosomal dyad, using two independent systems (White and Luger 2004; Edayathumangalam et al., unpublished results; Fig. 3A). In the first example (White and Luger 2004), we faithfully recreated a highly positioned nucleosome found in a particular eukaryotic promoter; the ligand in question is the DNA binding domain of the transcription factor Amt1 (Zhu and Thiele 1996). Using a combination of methods, such as gel electrophoresis, DNase footprinting, and fluorescence resonance energy transfer between defined regions within the nucleosome, we showed that the ends of the DNA partially dissociate upon binding of the factor near the nucleosomal dyad. Importantly, a full complement of histones is maintained in the ternary complex, and we have shown by fluorescence resonance energy transfer (a method that is highly sensitive to the relative distance of two fluorophores attached to H4 and H2B, respectively) that the H2A-H2B dimer and $(H3\text{-}H4)_2$ tetramer do not change their relative orientation.

Very similar results were found using a rather different experimental system. We have extensively used custom-made DNA minor groove binding ligands, the pyrrole-imidazole polyamides (Dervan and Edelson 2003) to probe the accessibility of nucleosomal DNA (Gottesfeld et al. 2001; Suto et al. 2003). As in the example described above, the binding site for one particular ligand was near the nucleosomal dyad. However, unlike Amt1, the polyamide could only access its binding site if either the H4 tail or the H3 tail is deleted (Gottesfeld et al. 2001). This finding suggests that simple steric occlusion of the binding site by the histone tails cannot be responsible for limiting accessibility. Indeed, structural analysis has later shown that the ends of the DNA locally unravel from the surface of the histone octamer, leading to changes in crystal packing (Edayathumangalam and Luger, manuscript in preparation). The energy barrier for this dynamic breathing of nucleosomal DNA appears to be lowered by the removal of the histone tails, and it is likely that less drastic tail modifications (such as phosphorylation or acetylation) may have similar (albeit more moderate) effects.

Together, these two results directly demonstrate the surprising adaptability of the nucleosome in response to DNA sequence and site-specific recognition, as has also been observed by the Widom laboratory (Li and Widom 2004; Li et al. 2005), and is supported by the finding that the ends of the DNA are held to the histone octamer by fewer hydrogen bonds than the remainder of the DNA (Luger and Richmond 1998a). The incorporation of histone variants with sequence variations in the DNA binding regions could also serve to modulate these breathing motions. In fact, one particular H2A histone variant (H2A.Bbd) appears to have evolved to weaken the contacts that stabilize the very end of nucleosomal DNA to the extent that only 118 base pairs are stably bound, as opposed to the canonical 147 base pairs (Bao et al. 2004).

2.3.2 Histone Dimer Removal and Exchange

The examples of nucleosome dynamics mentioned in the preceding section entail relatively simple breathing motions of the ends of nucleosomal DNA. The transient structural rearrangements are subtle and maintain almost complete DNA compaction; all histones remain stably bound within the nucleosome (White and Luger 2004). In order to release a more significant number of base pairs from the surface of the nucleosome in a more permanent manner, the entire H2A-H2B dimer would have to be removed from the nucleosome (Fig. 3B). Indeed, the transient removal of one or both H2A-H2B dimers from a nucleosome appears to be involved in many vital cellular processes (Baer and Rhodes 1983; Belotserkovskaya et al. 2003; Levchenko and Jackson 2004; Kireeva et al. 2002). Removal of the H2A-H2B dimers would free up approximately 40 base pairs of DNA on either side of the nucleosome and make these available for molecular recognition, while at the same time maintaining some degree of DNA compaction due to the remaining interactions of the $(H3\text{-}H4)_2$ tetramer with the DNA.

A second important implication for the transient removal of histone H2A-H2B dimers from chromatin is that this mechanism presents an opportunity to incorporate histone H2A variants into chromatin in a replication-independent manner (Fig. 3B). Substitution of one or more of the core histones with the corresponding histone variants has the potential to exert considerable influence on the structure and function of chromatin. Histone variants are distinct nonallelic forms of conventional, major-type histones that form the bulk of nucleosomes during replication and whose synthesis is tightly coupled to S-phase. They are characterized by a completely different expression pattern that is not restricted to S-phase. Histone variants are found in most eukaryotic organisms, and are expressed in all tissue types (unlike some H2B isoforms that are only found in specialized tissues such as testes). Intriguingly, the majority of histone variants have been identified for H2A (Chakravarthy et al. 2004). In one particular gene, the transcription-coupled replacement of H2A.Z containing histone dimers with major-type dimers appears to act as a cellular memory for transcriptionally poised gene domains (Farris et al. 2005). Specific chromatin remodeling factors appear to have evolved to promote the exchange of particular histone variants (reviewed in Korber and Horz 2004; Cairns 2005; see also Zhang et al. 2005; Walfridsson et al. 2005).

Transient removal of H2A-H2B dimers is facilitated by passing RNA polymerases (Farris et al. 2005; Levchenko et al. 2005; Kireeva et al. 2002) in conjunction with the FACT complex (Belotserkovskaya et al. 2003). A family of acidic proteins with an affinity for histones, the so-called histone chaperones (reviewed in Akey and Luger 2003) are also involved in this process (Levchenko and Jackson 2004). We have recently shown that one such chaperone, nucleosome assembly protein 1 (NAP-1) from yeast reversibly removes and replaces H2A-H2B or histone variant dimers from assembled nucleosomes in vitro, resulting in active histone exchange (Park et al. 2005). Since NAP-1 does not distinguish between H2A-H2B dimers containing major-type H2A or histone variants (Park et al. 2005), it is capable of replacing major-type H2A-H2B with variant dimers and vice versa. The removal of H2A-H2B dimers from nucleosomes requires an acidic region of NAP-1 that is not required for what has been assumed to be its premier function, chromatin assembly. It is possible that this particular region may help mimic DNA in the process of histone removal. We have also shown that the transient removal of one or both H2A-H2B dimers facilitates nucleosome sliding along the DNA to a thermodynamically favorable position, independently of ATP hydrolysis (Park et al. 2005). This indicates a more active role for NAP-1 in shaping chromatin structure than previously assumed. It remains to be seen whether the temperature-induced sliding observed on many DNA sequences in vitro and the ATP-dependent nucleosome sliding brought about by the large chromatin remodeling machines require the transient removal of the H2A-H2B dimer.

2.3.3 Complete Histone Removal

The most radical manifestation of dynamic nucleosomes is their complete removal (Fig. 3C). Although this was earlier deemed to be the only way by which eukaryotic DNA could be transcribed at all, this mechanism has until recently not received much attention in light of the emerging role of ATP-dependent chromatin remodeling factors that structurally change, but do not remove nucleosomes (Flaus and Owen-Hughes 2003b). There is now compelling evidence that nucleosomes (that is, all core histones) are lost completely from the PHO5 promoter upon activation in vivo (Boeger et al. 2003; Reinke and Horz 2003). The same two groups later showed that core histone loss from the promoter did not occur through sliding, but through transient eviction of all histones (Boeger et al. 2004; Korber et al. 2004). It now appears that the H3/H4 histone chaperone, Asf1, is either directly or indirectly responsible for histone removal on the PHO5 promoter (Adkins et al.

2004). Intriguingly, two recent genome-wide studies in yeast suggest that nucleosome dissociation from promoter regions might be a general phenomenon (Bernstein et al. 2004; Lee et al. 2004).

2.4 Summary and Outlook

The image of chromatin is changing from being viewed as a strict and rigid impediment to all cellular processes that require the DNA substrate, to a highly fluid and "shape-shifting" macromolecular assembly. In hindsight, this property should not have come as a surprise since the packaging of DNA into chromatin is obviously the primary determinant of its accessibility. New activities that impinge on chromatin structure are being discovered constantly, and there is no end in sight. The basic protein components of chromatin, the histones, appear to have evolved to confine DNA just enough to force it into forming a superhelix (resulting in an approximately fivefold compaction), but still allow the dynamic repositioning of the DNA on their surface or even their dynamic exchange. Additionally, the surface of the histone octamer that is not involved in DNA binding has the potential of forming docking surfaces for cellular factors, and of promoting nucleosome–nucleosome interactions to promote the formation of chromatin higher-order structure. Finally, the flexible histone tails are involved in a multitude of functions that are regulated by post-translational modifications and that involve the modulation of DNA accessibility within a mononucleosome and within the context of higher-order structure, as well as interactions with regulatory factors.

References

Adkins MW, Howar SR, Tyler JK (2004) Chromatin disassembly mediated by the histone chaperone Asf1 is essential for transcriptional activation of the yeast PHO5 and PHO8 genes. Mol Cell 14:657–666

Akey CW, Luger K (2003) Histone chaperones and nucleosome assembly. Curr Opin Struct Biol 13:6–14

Arents G, Moudrianakis EN (1995) The histone fold: a ubiquitous architectural motif utilized in DNA compaction and protein dimerization. Proc Natl Acad Sci U S A 92:11170–11174

Baer BW, Rhodes D (1983) Eukaryotic RNA polymerase II binds to nucleosome cores from transcribed genes. Nature 301:482–488

Bao Y, Konesky K, Park YJ, Rosu S, Dyer PN, Rangasamy D, Tremethick DJ, Laybourn PJ, Luger K (2004) Nucleosomes containing the histone variant H2ABbd organize only 118 base pairs of DNA. EMBO J 23:3314–3324

Belotserkovskaya R, Oh S, Bondarenko VA, Orphanides G, Studitsky VM, Reinberg D (2003) FACT facilitates transcription-dependent nucleosome alteration. Science 301:1090–1093

Berger SL (2002) Histone modifications in transcriptional regulation. Curr Opin Genet Dev 12:142–148

Bernstein BE, Liu CL, Humphrey EL, Perlstein EO, Schreiber SL (2004) Global nucleosome occupancy in yeast. Genome Biol 5:R62

Boeger H, Griesenbeck J, Strattan JS, Kornberg RD (2003) Nucleosomes unfold completely at a transcriptionally active promoter. Mol Cell 11:1587–1598

Boeger H, Griesenbeck J, Strattan JS, Kornberg RD (2004) Removal of promoter nucleosomes by disassembly rather than sliding in vivo. Mol Cell 14:667–673

Cairns BR (2005) Chromatin remodeling complexes: strength in diversity, precision through specialization. Curr Opin Genet Dev 15:185–190

Chakravarthy S, Bao Y, Roberts VA, Tremethick D, Luger K (2004) Structural characterisation of histone H2A variants. Cold Spring Harb Symp Quant Biol 69:227–234

Davey CA, Sargent DF, Luger K, Maeder AW, Richmond TJ (2002) Solvent mediated interactions in the structure of the nucleosome core particle at 1.9: a resolution. J Mol Biol 319:1097–1113

Dervan PB, Edelson BS (2003) Recognition of the DNA minor groove by pyrrole-imidazole polyamides. Curr Opin Struct Biol 13:284–299

Dorigo B, Schalch T, Bystricky K, Richmond TJ (2003) Chromatin fiber folding: requirement for the histone H4 N-terminal tail. J Mol Biol 327:85–96

Dorigo B, Schalch T, Kulangara A, Duda S, Schroeder RR, Richmond TJ (2004) Nucleosome arrays reveal the two-start organization of the chromatin fiber. Science 306:1571–1573

Edayathumangalam RS, Weyermann P, Dervan PB, Gottesfeld JM, Luger K (2005) Nucleosomes in solution exist as a mixture of twist-defect states. J Mol Biol 345:103–114

Farris SD, Rubio ED, Moon JJ, Gombert WM, Nelson BH, Krumm A (2005) Transcription-induced chromatin remodeling at the c-myc gene involves the local exchange of histone H2A.Z. J Biol Chem 280:25298–25303

Flaus A, Owen-Hughes T (2003a) Dynamic properties of nucleosomes during thermal and ATP-driven mobilization. Mol Cell Biol 23:7767–7779

Flaus A, Owen-Hughes T (2003b) Mechanisms for nucleosome mobilization. Biopolymers 68:563–578

Flaus A, Owen-Hughes T (2004) Mechanisms for ATP-dependent chromatin remodelling: farewell to the tuna-can octamer? Curr Opin Genet Dev 14:165–173

Gottesfeld JM, Melander C, Suto RK, Raviol H, Luger K, Dervan PB (2001) Sequence-specific recognition of DNA in the nucleosome by pyrrole-imidazole polyamides. J Mol Biol 309:625–639

Hansen JC (2002) Conformational dynamics of the chromatin fiber in solution: determinants mechanisms Functions. Annu Rev Biophys Biomol Struct 31:361–392

Jackson V (1990) In vivo studies on the dynamics of histone-DNA interaction: evidence for nucleosome dissolution during replication and transcription and a low level of dissolution independent of both. Biochemistry 29:719–731

Jenuwein T, Allis CD (2001) Translating the histone code. Science 293:1074–1080

Kimura H, Cook PR (2001) Kinetics of core histones in living human cells: little exchange of H3 and H4 and some rapid exchange of H2B. J Cell Biol 153:1341–1353

Kireeva ML, Walter W, Tchernajenko V, Bondarenko V, Kashlev M, Studitsky VM (2002) Nucleosome remodeling induced by RNAPolymerase II Loss of the H2A/H2B dimer during transcription. Mol Cell 9:541–552

Korber P, Horz W (2004) SWRred not shaken: mixing the histones. Cell 117:5–7

Korber P, Luckenbach T, Blaschke D, Horz W (2004) Evidence for histone eviction in trans upon induction of the yeast PHO5 promoter. Mol Cell Biol 24:10965–10974

Kornberg RD, Lorch Y (1991) Irresistible force meets immovable object: transcription and the nucleosome. Cell 67:833–836

Lee CK, Shibata Y, Rao B, Strahl BD, Lieb JD (2004) Evidence for nucleosome depletion at active regulatory regions genome-wide. Nat Genet 36:900–905

Levchenko V, Jackson V (2004) Histone release during transcription: NAP1 forms a complex with H2A and H2B and facilitates a topologically dependent release of H3 and H4 from the nucleosome. Biochemistry 43:2359–2372

Levchenko V, Jackson B, Jackson V (2005) Histone release during transcription: displacement of the two H2A-H2B dimers in the nucleosome is dependent on different levels of transcription-induced positive stress. Biochemistry 44:5357–5372

Li G, Widom J (2004) Nucleosomes facilitate their own invasion. Nat Struct Mol Biol 11:763–769

Li G, Levitus M, Bustamante C, Widom J (2005) Rapid spontaneous accessibility of nucleosomal DNA. Nat Struct Mol Biol 12:46–53

Luger K (2003) Structure and dynamic behavior of nucleosomes. Curr Opin Genet Dev 13:127–135

Luger K, Hansen JC (2005) Nucleosome and chromatin fiber dynamics. Curr Opin Struct Biol 15:188–196

Luger K, Richmond TJ (1998a) DNA binding within the nucleosome core. Curr Opin Struc Biol 8:33–40

Luger K, Richmond TJ (1998b) The histone tails of the nucleosome. Curr Opin Genet Dev 8:140–146

Luger K, Mader AW, Richmond RK, Sargent DF, Richmond TJ (1997a) Crystal structure of the nucleosome core particle at 2.8 Å resolution. Nature 389:251–259

Luger K, Rechsteiner TJ, Flaus AJ, Waye MM, Richmond TJ (1997b) Characterization of nucleosome core particles containing histone proteins made in bacteria. J Mol Biol 272:301–311

Park YJ, Dyer PN, Tremethick DJ, Luger K (2004) A new fluorescence resonance energy transfer approach demonstrates that the histone variant H2AZ stabilizes the histone octamer within the nucleosome. J Biol Chem 279:24274–24282

Park YJ, Chodaparambil JV, Bao Y, McBryant SJ, Luger K (2005) Nucleosome assembly protein 1 exchanges histone H2A-H2B dimers and assists nucleosome sliding. J Biol Chem 280:1817–1825

Polach KJ, Widom J (1995) Mechanism of protein access to specific DNA sequences in chromatin: a dynamic equilibrium model for gene regulation. J Mol Biol 254:130–149

Polach KJ, Widom J (1996) A model for the cooperative binding of eukaryotic regulatory proteins to nucleosomal target sites. J Mol Biol 258:800–812

Reinke H, Horz W (2003) Histones are first hyperacetylated and then lose contact with the activated PHO5 promoter. Mol Cell 11:1599–1607

Richmond TJ, Davey CA (2003) The structure of DNA in the nucleosome core. Nature 423:145–150

Strahl BD, Allis CD (2000) The language of covalent histone modifications. Nature 403:41–45

Sullivan S, Sink DW, Trout KL, Makalowska I, Taylor PM, Baxevanis AD, Landsman D (2002) The histone database. Nucleic Acids Res 30:341–342

Suto RK, Edayathumangalam RS, White CL, Melander C, Gottesfeld JM, Dervan PB, Luger K (2003) Crystal structures of nucleosome core particles in complex with minor groove DNA-binding ligands. J Mol Biol 326:371–380

Tsunaka Y, Kajimura N, Tate S, Morikawa K (2005) Alteration of the nucleosomal DNA path in the crystal structure of a human nucleosome core particle. Nucleic Acids Res 33:3424–3434

Van Holde KE (1988) Chromatin. Springer-Verlag, Berlin Heidelberg New York

Walfridsson J, Bjerling P, Thalen M, Yoo EJ, Park SD, Ekwall K (2005) The CHD remodeling factor Hrp1 stimulates CENP-A loading to centromeres. Nucleic Acids Res 33:2868–2879

White CL, Luger K (2004) Defined structural changes occur in a nucleosome upon Amt1 transcription factor binding. J Mol Biol 342:1391–1402

Whitlock JP Jr, Stein A (1978) Folding of DNA by histones which lack their NH2-terminal regions. J Biol Chem 253:3857–3861

Zhang R, Poustovoitov MV, Ye X, Santos HA, Chen W, Daganzo SM, Erzberger JP, Serebriiskii IG, Canutescu AA, Dunbrack RL et al. (2005) Formation of Macro H2A-containing senescence-associated heterochromatin foci and senescence driven by ASF1a and HIRA. Dev Cell 8:19–30

Zhu Z, Thiele DJ (1996) A specialized nucleosome modulates transcription factor access to a C glabrata metal responsive promoter. Cell 87:459–470

3 The Role of Snf2-Related Proteins in Cancer

T. Owen-Hughes

Abstract. Several HDAC inhibitors that exhibit impressive anti-tumour activity are now in clinical trials. Proteins that function in the same pathways might also serve as valuable therapeutic targets. A subset of histone deacetylase activities are found to be physically associated with ATP-dependent remodelling enzymes and may assist their function. This raises the possibility that ATP-dependent remodelling enzymes should be considered as therapeutic targets. Here some of the links between ATP-dependent chromatin remodelling enzymes and cancer are reviewed.

3.1 Introduction

The genomes of eukaryotes exist not as free DNA, but are associated with protein to form a complex termed chromatin. The fundamental subunit of chromatin is the nucleosome, which consists of an octamer of the four core histones around which DNA is wrapped in 1.6 negative superhelical turns (Luger et al. 1997). The assembly of DNA into chromatin contributes to the compaction of eukaryotic genomes required to physically package them within the confines of nuclei. Although one function of chromatin is to condense DNA, it is important that regulatory elements remain accessible.

In order to meet these conflicting requirements for condensation and accessibility, all eukaryotes have developed a specialized machinery capable of manipulating the chromatin environment so that it facilitates appropriate gene regulation. This machinery includes the use of enzymes that methylate DNA (Hermann et al. 2004), the use of proteins that bind to nucleosomes such as the linker histone H1 (Hansen 2002), and enzymes that alter the histone proteins by post-translational modification (Fischle et al. 2003).

The histone proteins are subject to a range of post-translational modifications, including acetylation, methylation, phosphorylation, ubiquitination, ribosylation (Zhang et al. 2003). Of these modifications, histone acetylation has probably been studied in most detail. For example, at the β-globin locus it has been recognized that changes in histone acetylation correlate with an increase in the sensitivity of the locus to be digested with DNaseI and the propensity for transcription in erythroid cells (Hebbes et al. 1994). In many cases, alterations in histone modification have been found to correlate with gene regulation, and enzymes responsible for modifications are intimately involved in gene regulation. This gives rise to the concept that histone modifications may comprise a code that adds content to the underlying genomic information (Jenuwein and Allis 2001). Most of the histone modifications studied to date are reversible, and in the case of histone acetylation, histone deacetylases (HDACs) are responsible for this. Interest in HDACs has grown following the use of HDAC inhibitors including butyric acid derivatives, depsipeptide, *N*-acetyl dinaline, pyroxamide, SAHA and valproic acid in clinical trials for cancer therapy (La Thangue 2004). All eukaryotes contain multiple HDACs belong to one of three major subfamilies (de Ruijter et al. 2003; Denu 2003) and there is good evidence that they function to regulated gene expression (Kurdistani and Grunstein 2003).

In many cases, alterations to histone acetylation occur as stages in pathways that also involve the action of ATP-dependent chromatin remodelling enzymes. In some cases, enzymes responsible for histone acetylation (Pray-Grant et al. 2005) or deacetylation (Xue et al. 1998; Zhang et al. 1999; Sif et al. 2001; Goldmark et al. 2000) are physically associated with ATP-dependent chromatin remodelling enzymes. This raises the possibility that ATP-dependent chromatin remodelling enzymes might also be valid therapeutic targets.

The Snf2 proteins can be classified into distinct subfamilies. Here some of the links between four of the subfamilies most closely linked to cancer are reviewed.

3.2 The SNF2 Subfamily

The human BRG1 and BRM proteins are homologues of the yeast Snf2 protein, the catalytic subunit of the SWI/SNF complex. While BRM knock-out mice are viable, BRG1 deletion results in embryonic lethality (Muchardt and Yaniv 2001). Interestingly, mice that are heterozygous for *BRG1* are susceptible to neoplasia and display large subcutaneous tumours (Bultman et al. 2000). There may be some redundancy in the function of BRG1 and BRM, but each also has unique functions. For example, BRG1 is expressed to higher levels in rapidly dividing cells (Reisman et al. 2005). Although BRG1 is required for development, both proteins can be depleted in some cell types. In these cases, it is possible that there is redundancy with other chromatin remodelling proteins.

It is likely that the majority of the BRG1 and BRM proteins exist as components of multi-subunit complexes closely related to the yeast SWI/SNF and RSC complexes (Wang et al. 1996; Xue et al. 2000). In yeast, mutation of different subunits confers phenotypes similar to that of the catalytic subunit (Peterson and Herskowitz 1992), and similar phenotypes are also observed in mammalian cells. For example, mice heterozygous for SNF5/INI1 develop nervous system and soft tissue sarcomas which may be related to the phenotype of BRG1 heterozygous mice (Klochendler-Yeivin et al. 2000). However, SNF5 mutation has a more dramatic phenotype than BRG1 mutation, raising the possibility that SNF5 may have some functions independent of BRG1.

Table 1. Links between human SWI/SNF complexes and cancer

Type of cancer	Link to human SWI/SNF	References
Malignant rhabdoid tumours	Truncating mutations of the hSNF5/INI1	Versteege et al. 1998
Chronic myeloid leukaemia	Deletions of the hSNF5/INI1 gene	Grand et al. 1999
Atypical teratoid/rhabdoid tumours of the brain and kidney	Mutations of the hSNF5/INI1	Biegel et al. 1999, 2000
Medulloblastoma and primitive neuroectodermal tumours	Deletion and mutations of the INI gene	Biegel et al. 2000
Acute leukaemia	The human SWI/SNF subunit ENL can be fused to the transcription factor MLL	Cairns 2001; Nie et al. 2003
Choroid plexus carcinomas and in a subset of central primitive neuroectodermal tumours and medulloblastomas	hSNF5/INI1 mutations	Sevenet et al. 1999a
Rhabdoid predisposition syndrome	hSNF5/INI1 mutations	Sevenet et al. 1999b
Rhabdoid tumours	Alteration of BAF47 expression	DeCristofaro et al. 1999
Breast, lung, pancreas, and prostate carcinomas	*BRG1* mutations identified in breast tumour cell lines	Wong et al. 2000
Prostate cancer	Germ-line single nucleotide polymorphisms in CpG islands of the *BRG1* gene have been linked to prostate cancer	Valdman et al. 2003
Lung cancers	BRG1 and BRM are lost in approximately 30% of human non-small cell lung cancers	Reisman et al. 2003; Fukuoka et al. 2004; Medina et al. 2004
Oral cancers	BRG1 mutations	Gunduz et al. 2005

In addition to associating with human homologues of SWI/SNF and RSC complex components, both BRG1 and BRM are components of several different multi-proteins complexes that be unique to higher eukaryotes. Many of the proteins that they associate with have links to cancer. For example, the association of BRG1 with Rb (Dunaief et al. 1994; He et al. 2003; Hendricks et al. 2004; Strobeck et al. 2000; Zhang et al. 2000), Lkb1 (Marignani et al. 2001), HDACs (Sif et al. 2001), nuclear receptors (Hsiao et al. 2003) and BRACA1 (Bochar et al. 2000b) may act to regulate cell cycle progression. However, one recent study suggests that human SWI/SNF function may be more important for differentiation than cell cycle arrest (de la Serna et al. 2001). For a recent review see Klochendler-Yeivin et al. (2002). The different ways in which human forms of the SWI/SNF complex function provide the basis for a range of links to a variety of human cancers. These are summarized in Table 1.

3.3 Mi2/CHD3/4 Subfamily

The Mi-2–NuRD complex contains both histone deacetylase and chromatin remodelling ATPase activities (Tong et al. 1998; Wade et al. 1998; Xue et al. 1998; Zhang et al. 1998). It is likely that this complex functions predominantly in the repression of transcription in pathways more directly related to histone deacetylases. Additional subunits of the Mi-2–NuRD complex include MTA1, MTA2 and MTA3. The founding member of the MTA family of proteins, MTA1, was identified in a screen designed to identify mRNAs up-regulated in metastatic breast cancer cells (Fujita et al. 2003; Pencil et al. 1993; Toh et al. 1994).

The MTA3-containing Mi-2–NuRD complex is thought to function downstream of oestrogen receptor to repress transcription of Snail. Expression of Snail causes repression of E-cadherin and epithelial to mesenchymal transitions (Fujita et al. 2003). Consistent with this, expression of the MTA1 gene is closely related to invasiveness and metastasis in non-small cell lung cancer (Sasaki et al. 2002), oesophageal cancers (Toh et al. 1999), thymoma (Sasaki et al. 2001), pancreatic cancer (Iguchi et al. 2000) and gastrointestinal carcinoma (Toh et al. 1997).

Table 2. Functions of human ISWI complexes

Complex	Possible function	Reference
CHRAC	DNA replication through heterochromatin	Collins et al. 2002
WICH	Maintenance of chromatin structures during DNA replication	Poot et al. 2004
NURF	Regulation of transcription, possible link to leukaemia, involved in maintaining higher-order chromatin structure	Badenhorst et al. 2002; Barak et al. 2003
RSF	Possible role in transcription through chromatin	LeRoy et al. 1998; Loyola et al. 2003
WCRF	Not yet known	Bochar et al. 2000a
mNoRC	Regulation of rDNA transcription	Strohner et al. 2004
SNF2h/cohesin	Cohesin loadin	Hakimi et al. 2002

3.4 The ISWI Subfamily

The human homologues of the ISWI proteins are found as components of multiple complexes (Table 2). These complexes have been proposed to have roles in a variety of processes, including the activation and repression of transcription, DNA replication and the maintenance of chromosome structure. As characterization of these complexes is not yet complete, it is likely that additional functions remain to be found. While many of the known functions provide potential links to cancer, the evidence for a direct involvement in cancer progression is not as strong as for the Snf2 family. This may be due to the fact that the role of these complexes in cancer has not been investigated sufficiently.

3.5 The LSH Subfamily

Action of the Arabidopsis homologue of LSH, DDM1, has been found to be required for DNA methylation (Bourc'his and Bestor 2002). Alterations to DNA methylation status have been extensively correlated with the progression towards cancer (Ushijima 2005) and so implicate LSH proteins in this process.

Mutations in LSH, which is also known as PASG and HELLS, have been identified in approximately 50% of myelogenous and lymphoblastic leukaemia (Lee et al. 2000). LSH null mice do not survive to an age at which tumours would be expected; however, DNA methylation is reduced (Geiman et al. 2001).

3.6 Conclusions

The number of links between ATP-dependent chromatin remodelling enzymes and cancer are increasing rapidly. Some of these links are likely to represent alterations that occur as a result of the progression towards cancer rather than being causal. In addition, different remodelling enzymes play distinct roles in cancer. For example the BRG1 and BRM proteins appear to function primarily as tumour suppressors whereas components of the Mi-2 complex are overexpressed in cancer cells. This raises the possibility that therapeutic strategies would require specificity for specific Snf2 subfamilies. However, based on the paradigm of the HDACs, which also have diverse functions, this may not be essential. While this review has concentrated on the four subfamilies with strongest links to cancer, it is likely that once other subfamilies have been better characterized they will provide further links. In this respect, ATP-dependent remodelling enzymes that participate in DNA repair may be of special interest.

Acknowledgements. Thanks to members of the TOH lab for comments on the manuscript. This work was supported by a Wellcome Trust Senior Fellowship.

References

Badenhorst P, Voas M, Rebay I, Wu C (2002) Biological functions of the ISWI chromatin remodeling complex NURF. Genes Dev 16:3186–3198

Barak O, Lazzaro MA, Lane WS, Speicher DW, Picketts DJ, Shiekhattar R (2003) Isolation of human NURF: a regulator of Engrailed gene expression. EMBO J 22:6089–6100

Biegel JA, Zhou JY, Rorke LB, Stenstrom C, Wainwright LM, Fogelgren B (1999) Germline and acquired mutations of INI1 in atypical teratoid and rhabdoid tumors. Cancer Res 59:74–79

Biegel JA, Fogelgren B, Wainwright LM, Zhou JY, Bevan H, Rorke LB (2000) Germline INI1 mutation in a patient with a central nervous system atypical teratoid tumor and renal rhabdoid tumor. Genes Chromosom Cancer 28:31–37

Bochar DA, Savard J, Wang W, Lafleur DW, Moore P, Cote J, Shiekhattar R (2000a) A family of chromatin remodeling factors related to Williams syndrome transcription factor. Proc Natl Acad Sci U S A 97:1038–1043

Bochar DA, Wang L, Beniya H, Kinev A, Xue Y, Lane WS, Wang W, Kashanchi F, Shiekhattar R (2000b) BRCA1 is associated with a human SWI/SNF-related complex: linking chromatin remodeling to breast cancer. Cell 102:257–265

Bourc'his D, Bestor TH (2002) Helicase homologues maintain cytosine methylation in plants and mammals. Bioessays 24:297–299

Bultman S, Gebuhr T, Yee D, La Mantia C, Nicholson J, Gilliam A, Randazzo F, Metzger D, Chambon P, Crabtree G, Magnuson T (2000) A Brg1 null mutation in the mouse reveals functional differences among mammalian SWI/SNF complexes. Mol Cell 6:1287–1295

Cairns BR (2001) Emerging roles for chromatin remodeling in cancer biology. Trends Cell Biol 11:S15–S21

Collins N, Poot RA, Kukimoto I, Garcia-Jimenez C, Dellaire G, Varga-Weisz PD (2002) An ACF1-ISWI chromatin-remodeling complex is required for DNA replication through heterochromatin. Nat Genet 32:627–632

De la Serna IL, Roy K, Carlson KA, Imbalzano AN (2001) MyoD can induce cell cycle arrest but not muscle differentiation in the presence of dominant negative SWI/SNF chromatin remodeling enzymes. J Biol Chem 276:41486–41491

De Ruijter AJ, van Gennip AH, Caron HN, Kemp S, van Kuilenburg AB (2003) Histone deacetylases (HDACs): characterization of the classical HDAC family. Biochem J 370:737–749

De Cristofaro MF, Betz BL, Wang W, Weissman BE (1999) Alteration of hSNF5/INI1/BAF47 detected in rhabdoid cell lines and primary rhabdomyosarcomas but not Wilms' tumors. Oncogene 18:7559–7565

Denu JM (2003) Linking chromatin function with metabolic networks: Sir2 family of NAD(+)-dependent deacetylases. Trends Biochem Sci 28:41–48

Dunaief JL, Strober BE, Guha S, Khavari PA, Alin K, Luban J, Begemann M, Crabtree GR, Goff SP (1994) The retinoblastoma protein and BRG1 form a complex and cooperate to induce cell cycle arrest. Cell 79:119–130

Fischle W, Wang Y, Allis CD (2003) Histone and chromatin cross-talk. Curr Opin Cell Biol 15:172–183

Fujita N, Jaye DL, Kajita M, Geigerman C, Moreno CS, Wade PA (2003) MTA3, a Mi-2/NuRD complex subunit, regulates an invasive growth pathway in breast cancer. Cell 113:207–219

Fukuoka J, Fujii T, Shih JH, Dracheva T, Meerzaman D, Player A, Hong K, Settnek S, Gupta A, Buetow K et al. (2004) Chromatin remodeling factors and BRM/BRG1 expression as prognostic indicators in non-small cell lung cancer. Clin Cancer Res 10:4314–4324

Geiman TM, Tessarollo L, Anver MR, Kopp JB, Ward JM, Muegge K (2001) Lsh, a SNF2 family member, is required for normal murine development. Biochim Biophys Acta 1526:211–220

Goldmark JP, Fazzio TG, Estep PW, Church GM, Tsukiyama T (2000) The Isw2 chromatin remodeling complex represses early meiotic genes upon recruitment by Ume6p. Cell 103:423–433

Grand F, Kulkarni S, Chase A, Goldman JM, Gordon M, Cross NC (1999) Frequent deletion of hSNF5/INI1, a component of the SWI/SNF complex, in chronic myeloid leukemia. Cancer Res 59:3870–3874

Gunduz E, Gunduz M, Ouchida M, Nagatsuka H, Beder L, Tsujigiwa H, Fukushima K, Nishizaki K, Shimizu K, Nagai N (2005) Genetic and epigenetic alterations of BRG1 promote oral cancer development. Int J Oncol 26:201–210

Hakimi MA, Bochar DA, Schmiesing JA, Dong Y, Barak OG, Speicher DW, Yokomori K, Shiekhattar R (2002) A chromatin remodelling complex that loads cohesin onto human chromosomes. Nature 418:994–998

Hansen JC (2002) Conformational dynamics of the chromatin fiber in solution: determinants, mechanisms, and functions. Annu Rev Biophys Biomol Struct 31:361–392

He S, Bauman D, Davis JS, Loyola A, Nishioka K, Gronlund JL, Reinberg D, Meng F, Kelleher N, McCafferty DG (2003) Facile synthesis of site-specifically acetylated and methylated histone proteins: reagents for evaluation of the histone code hypothesis. Proc Natl Acad Sci U S A 100:12033–12038

Hebbes TR, Clayton AL, Thorne AW, Crane-Robinson C (1994) Core histone hyperacetylation co-maps with generalized DNase I sensitivity in the chicken β-globin chromosomal domain. EMBO J 13:1823–1830

Hendricks KB, Shanahan F, Lees E (2004) Role for BRG1 in cell cycle control and tumor suppression. Mol Cell Biol 24:362–376

Hermann A, Gowher H, Jeltsch A (2004) Biochemistry and biology of mammalian DNA methyltransferases. Cell Mol Life Sci 61:2571–2587

Hsiao PW, Fryer CJ, Trotter KW, Wang W, Archer TK (2003) BAF60a mediates critical interactions between nuclear receptors and the BRG1 chromatin-remodeling complex for transactivation. Mol Cell Biol 23:6210–6220

Iguchi H, Imura G, Toh Y, Ogata Y (2000) Expression of MTA1, a metastasis-associated gene with histone deacetylase activity in pancreatic cancer. Int J Oncol 16:1211–1214

Jenuwein T, Allis CD (2001) Translating the histone code. Science 293:1074–1080

Klochendler-Yeivin A, Fiette L, Barra J, Muchardt C, Babinet C, Yaniv M (2000) The murine SNF5/INI1 chromatin remodeling factor is essential for embryonic development and tumor suppression. EMBO Rep 1:500–506

Klochendler-Yeivin A, Muchardt C, Yaniv M (2002) SWI/SNF chromatin remodeling and cancer. Curr Opin Genet Dev 12:73–79

Kurdistani SK, Grunstein M (2003) Histone acetylation and deacetylation in yeast. Nat Rev Mol Cell Biol 4:276–284

La Thangue NB (2004) Histone deacetylase inhibitors and cancer therapy. J Chemother 16 [Suppl 4]:64–67

Lee DW, Zhang K, Ning ZQ, Raabe EH, Tintner S, Wieland R, Wilkins BJ, Kim JM, Blough RI, Arceci RJ (2000) Proliferation-associated SNF2-like gene (PASG): a SNF2 family member altered in leukemia. Cancer Res 60:3612–3622

LeRoy G, Orphanides G, Lane WS, Reinberg D (1998) Requirement of RSF and FACT for transcription of chromatin templates in vitro. Science 282:1900–1904

Loyola A, Huang JY, LeRoy G, Hu S, Wang YH, Donnelly RJ, Lane WS, Lee SC, Reinberg D (2003) Functional analysis of the subunits of the chromatin assembly factor RSF. Mol Cell Biol 23:6759–6768

Luger K, Mader AW, Richmond RK, Sargent DF, Richmond TJ (1997) Crystal structure of the nucleosome core particle at 2.8 A resolution. Nature 389:251–260

Marignani PA, Kanai F, Carpenter CL (2001) LKB1 associates with Brg1 and is necessary for Brg1-induced growth arrest. J Biol Chem 276:32415–32418

Medina PP, Carretero J, Fraga MF, Esteller M, Sidransky D, Sanchez-Cespedes M (2004) Genetic and epigenetic screening for gene alterations of the chromatin-remodeling factor SMARCA4/BRG1, in lung tumors. Genes Chromosom Cancer 41:170–177

Muchardt C, Yaniv M (2001) When the SWI/SNF complex remodels. . . the cell cycle. Oncogene 20:3067–3075

Nie Z, Yan Z, Chen EH, Sechi S, Ling C, Zhou S, Xue Y, Yang D, Murray D, Kanakubo E et al. (2003) Novel SWI/SNF chromatin-remodeling complexes contain a mixed-lineage leukemia chromosomal translocation partner. Mol Cell Biol 23:2942–2952

Pencil SD, Toh Y, Nicolson GL (1993) Candidate metastasis-associated genes of the rat 13762NF mammary adenocarcinoma. Breast Cancer Res Treat 25:165–174

Peterson CL, Herskowitz I (1992) Characterization of the yeast SWI1, SWI2, SWI3 genes which encode a global activator of transcription. Cell 68:573–583

Poot RA, Bozhenok L, van den Berg DL, Steffensen S, Ferreira F, Grimaldi M, Gilbert N, Ferreira J, Varga-Weisz PD (2004) The Williams syndrome transcription factor interacts with PCNA to target chromatin remodelling by ISWI to replication foci. Nat Cell Biol 6:1236–1244

Pray-Grant MG, Daniel JA, Schieltz D, Yates JR 3rd, Grant PA (2005) Chd1 chromodomain links histone H3 methylation with SAGA- and SLIK-dependent acetylation. Nature 433:434–438

Reisman DN, Sciarrotta J, Wang W, Funkhouser WK, Weissman BE (2003) Loss of BRG1/BRM in human lung cancer cell lines and primary lung cancers: correlation with poor prognosis. Cancer Res 63:560–566

Reisman DN, Sciarrotta J, Bouldin TW, Weissman BE, Funkhouser WK (2005) The expression of the SWI/SNF ATPase subunits BRG1 and BRM in normal human tissues. Appl Immunohistochem Mol Morphol 13:66–74

Sasaki H, Yukiue H, Kobayashi Y, Nakashima Y, Kaji M, Fukai I, Kiriyama M, Yamakawa Y, Fujii Y (2001) Expression of the MTA1 mRNA in thymoma patients. Cancer Lett 174:159–163

Sasaki H, Moriyama S, Nakashima Y, Kobayashi Y, Yukiue H, Kaji M, Fukai I, Kiriyama M, Yamakawa Y, Fujii Y (2002) Expression of the MTA1 mRNA in advanced lung cancer. Lung Cancer 35:149–154

Sevenet N, Lellouch-Tubiana A, Schofield D, Hoang-Xuan K, Gessler M, Birnbaum D, Jeanpierre C, Jouvet A, Delattre O (1999a) Spectrum of hSNF5/INI1 somatic mutations in human cancer and genotype-phenotype correlations. Hum Mol Genet 8:2359–2368

Sevenet N, Sheridan E, Amram D, Schneider P, Handgretinger R, Delattre O (1999b) Constitutional mutations of the hSNF5/INI1 gene predispose to a variety of cancers. Am J Hum Genet 65:1342–1348

Sif S, Saurin AJ, Imbalzano AN, Kingston RE (2001) Purification and characterization of mSin3A-containing Brg1 and hBrm chromatin remodeling complexes. Genes Dev 15:603–618

Strobeck MW, Knudsen KE, Fribourg AF, DeCristofaro MF, Weissman BE, Imbalzano AN, Knudsen ES (2000) BRG-1 is required for RB-mediated cell cycle arrest. Proc Natl Acad Sci U S A 97:7748–7753

Strohner R, Nemeth A, Nightingale KP, Grummt I, Becker PB, Langst G (2004) Recruitment of the nucleolar remodeling complex NoRC establishes ribosomal DNA silencing in chromatin. Mol Cell Biol 24:1791–1798

Toh Y, Pencil SD, Nicolson GL (1994) A novel candidate metastasis-associated gene, mta1, differentially expressed in highly metastatic mammary adenocarcinoma cell lines. cDNA cloning, expression, and protein analyses. J Biol Chem 269:22958–22963

Toh Y, Oki E, Oda S, Tokunaga E, Ohno S, Maehara Y, Nicolson GL, Sugimachi K (1997) Overexpression of the MTA1 gene in gastrointestinal carcinomas: correlation with invasion and metastasis. Int J Cancer 74:459–463

Toh Y, Kuwano H, Mori M, Nicolson GL, Sugimachi K (1999) Overexpression of metastasis-associated MTA1 mRNA in invasive oesophageal carcinomas. Br J Cancer 79:1723–1726

Tong JK, Hassig CA, Schnitzler GR, Kingston RE, Schreiber SL (1998) Chromatin deacetylation by an ATP-dependent nucleosome remodelling complex. Nature 395:917–921

Ushijima T (2005) Detection and interpretation of altered methylation patterns in cancer cells. Nat Rev Cancer 5:223–231

Valdman A, Nordenskjold A, Fang X, Naito A, Al-Shukri S, Larsson C, Ekman P, Li C (2003) Mutation analysis of the BRG1 gene in prostate cancer clinical samples. Int J Oncol 22:1003–1007

Versteege I, Sevenet N, Lange J, Rousseau-Merck MF, Ambros P, Handgretinger R, Aurias A, Delattre O (1998) Truncating mutations of hSNF5/INI1 in aggressive paediatric cancer. Nature 394:203–206

Wade PA, Jones PL, Vermaak D, Wolffe AP (1998) A multiple subunit Mi-2 histone deacetylase from Xenopus laevis cofractionates with an associated Snf2 superfamily ATPase. Curr Biol 8:843–846

Wang W, Xue Y, Zhou S, Kuo A, Cairns BR, Crabtree GR (1996) Diversity and specialization of mammalian SWI/SNF complexes. Genes Dev 10:2117–2130

Wong AK, Shanahan F, Chen Y, Lian L, Ha P, Hendricks K, Ghaffari S, Iliev D, Penn B, Woodland AM et al. (2000) BRG1, a component of the SWI-SNF complex, is mutated in multiple human tumor cell lines. Cancer Res 60:6171–6177

Xue Y, Wong J, Moreno GT, Young MK, Cote J, Wang W (1998) NURD, a novel complex with both ATP-dependent chromatin-remodeling and histone deacetylase activities. Mol Cell 2:851–861

Xue Y, Canman JC, Lee CS, Nie Z, Yang D, Moreno GT, Young MK, Salmon ED, Wang W (2000) The human SWI/SNF-B chromatin-remodeling complex is related to yeast rsc and localizes at kinetochores of mitotic chromosomes. Proc Natl Acad Sci U S A 97:13015–13020

Zhang HS, Gavin M, Dahiya A, Postigo AA, Ma D, Luo RX, Harbour JW, Dean DC (2000) Exit from G1 and S phase of the cell cycle is regulated by repressor complexes containing HDAC-Rb-hSWI/SNF and Rb-hSWI/SNF. Cell 101:79–89

Zhang L, Eugeni EE, Parthun MR, Freitas MA (2003) Identification of novel histone post-translational modifications by peptide mass fingerprinting. Chromosoma 112:77–86

Zhang Y, LeRoy G, Seelig HP, Lane WS, Reinberg D (1998) The dermatomyositis-specific autoantigen Mi2 is a component of a complex containing histone deacetylase and nucleosome remodeling activities. Cell 95:279–289

Zhang Y, Ng HH, Erdjument-Bromage H, Tempst P, Bird A, Reinberg D (1999) Analysis of the NuRD subunits reveals a histone deacetylase core complex and a connection with DNA methylation. Genes Dev 13:1924–1935

4 Imitation Switch Complexes

J. Mellor

Abstract. The imitation switch (ISWI) family of chromatin remodelling ATPases is found in organisms ranging from yeast to mammals. ISWI ATPases assemble chromatin and slide and space nucleosomes, making the chromatin template fluid and allowing appropriate regulation of events such as transcription, DNA replication, recombination and repair. The site of action of the ATPases is determined, in part by the tissue type in which the enzyme is expressed and in part by the nature of the proteins associated with the enzyme. The ISWI complexes are generally conserved in composition and function across species. Roles in gene expression and DNA replication in heterochromatin, gene activation and repression in euchromatin, and functions related to maintaining chromosome architecture are associated with different complexes. Defects in ISWI-associated proteins may be associated with neurodegenerative disease, anencephaly, William's syndrome and melanotic tumours. Finally, the mechanism by which yeast Isw1b influences gene transcription is discussed.

4.1 Introduction

The imitation switch (ISWI) class of chromatin remodelling ATPase belongs to a larger family of chromatin remodelling ATPases that includes the SWI/SNF class, the Mi-2/CHD1 class and the INO80-related enzymes (Becker and Horz 2002). Chromatin remodelling enzymes use the energy of ATP to alter the structure or positioning of nucleosomes, the basic building blocks of chromatin, and so regulate access of key sites on the DNA and influence higher-order structures within chromosomes. Members of the ISWI class are found ubiquitously in eukaryotes, with the notable exception of the fission yeast *Schizosaccharomyces pombe*. The ISWI enzymes can assemble chromatin in vitro, can mobilize nucleosomes both in vitro and in vivo and are found in complexes with proteins that enhance and direct these activities. Moreover, other chromatin modifying or remodelling activities may reside within the complexes, which help to establish the outcome of the ISWI-determined nucleosome mobilization. Thus ISWI factors are associated with both repressing and activating events on chromatin and influence transcription, DNA replication and chromosome structure. Roles in DNA repair and recombination are also likely. The formation of complexes also appears to facilitate the recruitment of enzymes or factors that regulate events in transcription or DNA replication unrelated to chromatin modification or remodelling. Thus understanding the functions associated with ISWI complexes is both exciting and challenging. In this short overview, our current understanding of ISWI complexes in a variety of organisms, including roles in disease, development and differentiation, is discussed.

4.2 The ISWI Proteins

4.2.1 ISWI ATPases Contain Distinct Structural Motifs

All ISWI proteins share a common motif towards the C-terminus, the HAND-SANT domain (Boyer et al. 2002) linked to the SLIDE domain by a long alpha helical spacer (Grune et al. 2003). The ATPase domain is within the amino terminus of the protein. The SLIDE domain resembles a myb-like DNA binding domain and is a distinguishing and unique feature of all ISWI proteins (Fig. 1). The SANT domain, together with the HAND domain, is proposed to interact with histone tails. This may involve binding to the H3 or H4 tails to facilitate remodelling or presentation of the tails for subsequent modification by other modifying enzymes (see de la Cruz et al. 2005 for a recent review). This domain organization and proposed functions for the individual motifs are consistent with the contacts made on a nucleosomal substrate by an ISWI

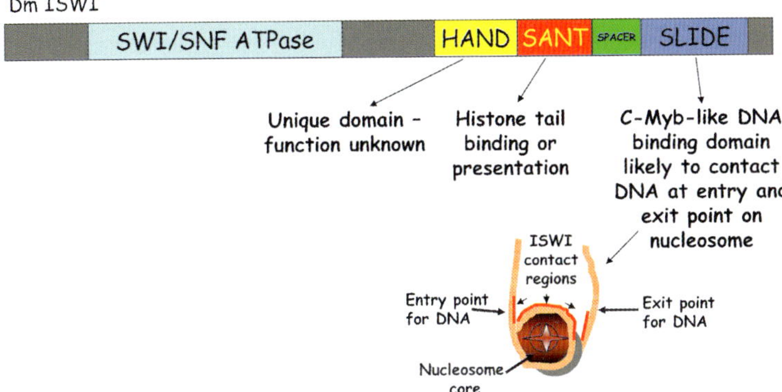

Fig. 1. Schematic showing the conserved domains on Dm ISWI protein. The data are taken from Grune et al. (2003). ISWI contacts nucleosomes and both the SANT and SLIDE domains are proposed to play a role. The SANT domain is associated with binding to, or presentation of, the histone tails on the nucleosomes. By contrast, the SLIDE domain is proposed to contact nucleosomal DNA directly through a myb-like DNA-binding domain. Mapping contact sites between DmISWI indicate contact points at the entry and exit points (Schwanbeck et al. 2004). These are shown on the schematic of a nucleosome core particle as *red lines* on the DNA strand

complex (Schwanbeck et al. 2004). Histone tails are required together with DNA at the entry/exit point on the nucleosomes.

4.2.2 Related ISWI ATPases in Eukaryotes

In some species, there are two highly related genes encoding ISWI proteins, in others only one. Budding yeast and mammals contain two proteins, Isw1 and Isw2 or SNF2h and SNF2L, respectively. Isw1 and Isw2 differ in their capacity to remodel nucleosomes (Tsukiyama et al. 1999; Vary et al. 2003). SNF2H and SNF2L show distinct tissue-specific activity (Barak et al. 2004). In mice, SNF2H is ubiquitously expressed, while the expression of SNF2L is limited to neuronal and gonadal tissues. By contrast, in humans, SNF2H and SNF2L are ubiquitously expressed. However, a spliced variant of SNF2L, known as +13, which allows the formation of complexes but inhibits remodelling activity, limits the activity of hSNF2L to neuronal tissues, like the mouse (Barak et al. 2004). Flies and frogs have one protein, known as ISWI, ubiquitously expressed. Despite these differences, there is a similar range of activities associated with the ATPases from yeast to humans, and this is likely to be determined by the nature and properties of the associated proteins.

4.3 ISWI Complexes

In budding yeast, a systematic analysis of the proteome using tandem affinity purification (TAP) tags has identified up to 13 complexes containing Isw1 and/or Isw2. As the composition and possible functions associated with these complexes has been reviewed recently, these will not be discussed here (Mellor and Morillon 2004). Instead, the focus will be on those ISWI containing complexes that have been biochemically characterized. The composition of these complexes and a summary of associated functions are summarized in Table 1. There are a number of unifying themes. First, many ISWI-associated proteins across the species contain conserved folds and domains. Acf1, WSTF, WCRF, TIP5, NURF301, BPTF, CERC-2, Ioc2 and Itc1 all share important functional domains that may influence the action of the protein on the ATPase activity of the ISWI protein. Second, many of the complexes

Table 1. Biochemically characterized complexes containing imitation switch (ISWI) remodelling activities

Organism	ISWI	Complex name	Associated proteins	Function	Reference
Yeast	Isw1	Isw1a	Ioc3	Not essential for viability Involved primarily in repression of gene expression Many complexes containing Isw1 and Ioc3 have been identified. Some also contain Isw2 and Itc1.	Cuperus and Shore 2002; Gavin 2002; Mellor and Morillon 2004; Morillon et al. 2003; Vary et al. 2003
		Isw1b	Ioc2 Ioc4	Not essential for viability. Involved in transcription elongation and termination	Morillon et al. 2003; Vary et al. 2003
	Isw2		Itc1 Dpb4 Dsl1	A homologue of DmCHRAC that antagonizes the repressive action of DNA Pol epsilon at telomeric heterochromatin Itc1 is a homologue of WSTF/ACF and likely to share function in counteracting the spread of heterochromatin.	(Iida and Araki 2004; McConnell et al. 2004
	Isw2		Itc1	Works cooperatively with Rpd3/SIN3 HDAC/corepressor and Ssn6/Tup1 co-repressors to repress gene expression at a number of loci. Isw2 may be recruited though DNA bound factors such as Ume6. A homologue of ACF?	Fazzio et al. 2001; Goldmark et al. 2000; Kent et al. 2001; Ruiz et al. 2003; Sugiyama and Nikawa 2001; Trachtulcova et al. 2000, 2004; Zhang and Reese 2004a, 2004b)
Drosophila	ISWI			*DmISWI* is essential for viability. Loss of ISWI function in vivo results in misregulation of transcription and global alterations in chromosome structure	Deuring et al. 2000

Table 1. (continued)

Organism	ISWI	Complex name	Associated proteins	Function	Reference
	ISWI	NURF	NURF301 NURF55 NURF38	NURF (nucleosome remodelling factor) is required for transcription activation in vivo at heat shock and homeotic genes. Mutants in NURF301 associated with melanotic tumours	Badenhorst et al. 2002; Hamiche et al. 1999; Martinez Balbas et al. 1998; Xiao et al. 2001
	ISWI	TRF-2	TRF-2 (TATA binding protein [TBP]-related factor 2). NURF-55, NURF-38. DREF (DNA replication-related element binding factor)	TRF2s are required for proper embryonic development and differentiation. TRF-2 coordinates transcription of a subset of genes including those encoding PCNA, DNA replication factors and cell proliferation factors. Acts as a core-promoter selectivity factor Note that NURF301 cannot be detected in this complex	Hochheimer et al. 2002
	ISWI	ACF	Acf1	*acf1* is not required for viability. Chromatin assembly activity. Gene repression and repression of DNA replication through formation of repressive chromatin including histone H1. Sliding enhanced by DNA chaperone HMGB1	Eberharter et al. 2001, 2004; Bonaldi 2002; Fyodorov et al. 2004; Lusser et al. 2005

Table 1. (continued)

Organism	ISWI	Complex name	Associated proteins	Function	Reference
	ISWI	CHRAC	Acf1, p15, p17	Enhances nucleosome sliding and assembly by ACF HMGB1 facilitates ACF/CHRAC-dependent nucleosome sliding.	Bonaldi et al. 2002; Kukimoto et al. 2004
Xenopus	ISWI	X-CHRAC X-WICH	Two complexes reported: ACF1 plus Two small subunits likely to be X-CHRAC and a second complex including X-WSTF	XISWI is a major component of mitotic chromosomes in *Xenopus* eggs. Association of XISWI is cell cycle regulated and under control of INCEPP-aurora B kinase (H3S10P) Essential for nucleosome spacing but not histone deposition, DNA replication, chromosome condensation or sister chromatid cohesion. Note the egg chromosomes are profoundly different from those in somatic cells and do not have extensive regions of heterochromatin as founding somatic chromosomes. A possible role in TBP displacement reported for XISWI	Bozhenok et al. 2002; Kikyo et al. 2000; MacCallum et al. 2002
Mammals		SNF2H		Snf2h is essential for viability and has roles in differentiation and development	Stopka and Skoultchi 2003
Mammals	SNF2h	WCRF/ hACF	WSTF-related chromatin remodelling factor WCRF180 – related to WSTF and WCRF135 – SNF2h	Required for DNA replication through heterochromatin. Does not form stable association with mitotic chromosomes (cf WICH)	Bochar et al. 2000; Collins et al. 2002; LeRoy et al. 2000

Table 1. (continued)

Organism	ISWI	Complex name	Associated proteins	Function	Reference
	SNF2h	WICH	WSTF-ISWI chromatin-remodelling complex	WSTF interacts with PCNA at replication foci. Required for DNA replication through heterochromatin WSTF, binds stably to mitotic chromosomes	Bozhenok et al. 2002; Poot et al. 2004
	SNF2h	RSF	Remodelling and spacing factor containing p325 (Rsf-1)	Required for nucleosome deposition and regularly spaced nucleosome arrays	LeRoy et al. 1998; Loyola et al. 2003
	SNF2h	NoRC	Nucleosome-remodelling complex containing TIP5 (TTF-I-interacting protein 5)	Nucleolar factors that establish heterochromatic structures at rDNA promoters and controls replication timing of rDNA	Li et al. 2005; Strohner et al. 2001
	SNF2h	huCHRAC	ACF CHRAC-15 CHRAC-17	Chromatin assembly not sliding by ACF is enhanced by histone fold proteins	Kukimoto et al. 2004

Table 1. (continued)

Organism	ISWI	Complex name	Associated proteins	Function	Reference
	SNF2h	SNF2H-Cohesin	SMC1, SMC3, SA1/SA2, hRAD21 plus components of NuRD (Mi2 + RbAp46.48 + HDAC2)	May recruit cohesin to regions of chromatin with specifically modified histone tails containing Alu repeats. Binding is also influenced by state of DNA methylation. SNF2h binds directly to hRAD21 of the cohesin complex	Hakimi et al. 2002
	SNF2L	NURF	BPTF (orthologue of DmNURF301) RbAP48/46 (orthologue of DmNURF 55) Lacks equivalent to DmNURF38	Expression is limited to neurons and a few other tissues. Role in brain development. Regulates *engrailed* expression	Barak et al. 2003
	SNF2L	CERF	CECR-2	Role in neurulation	Banting et al. 2005
	SNF2L	SNF2L+13		Dominant negative splice variant form of SNF2L – expressed ubiquitously in human, not mice, except in neural tissues. Forms complexes but has no remodelling activity	Barak et al. 2004

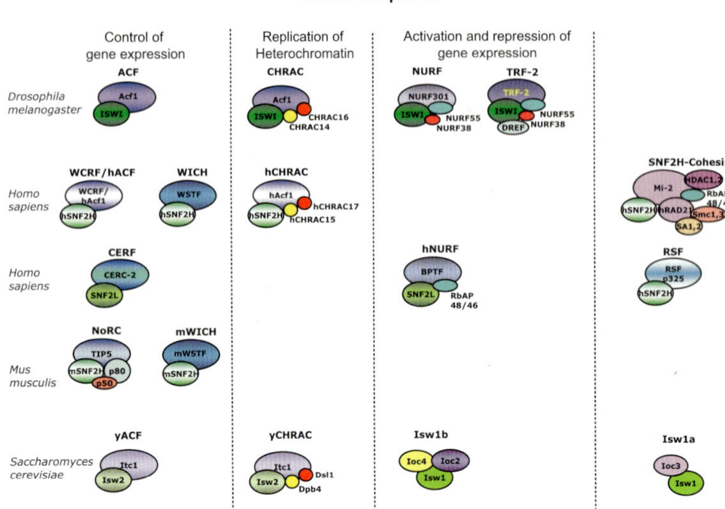

Fig. 2. Schematic showing the composition of ISWI complexes biochemically isolated from the species shown in the left-hand column. The name of the complex is indicated above each representation and the identity of each component indicated next to or within the *ovals* that represent the subunits. The schematic is roughly divided into four functional classes, separated by *broken lines*: the ACF-related complexes that control gene expression, the ACF-related complexes with roles in the replication of heterochromatin, the NURF-related complexes with roles in gene expression and the complexes with functions unrelated to the first three classes

have conserved composition across species (Fig. 2). Finally, many of the functions associated with ISWI complexes are broadly conserved across species.

4.3.1 ACF and Related Complexes: CHRAC, WSTF and NoRC

A conserved activity, associated with budding yeast, flies, frogs and mammals, relates to the formation of, and regulated replication and transcription of, repressive chromatin including heterochromatin. Formation of repressive chromatin is an activity that is clearly associated with ACF1 (ATP-utilizing chromatin assembly and remodelling factors) in flies, frogs and human and the related Isw2–Itc1 complex in budding yeast (Collins et al. 2002; Iida and Araki 2004; LeRoy et al. 2000; Lusser et al. 2005; Ruiz et al. 2003; Sugiyama and Nikawa 2001; Trachtulcova et al. 2004). Functionally, most is known about the yeast enzyme, yACF, and its role in repressing gene expression. First, Isw2–Itc1 may be targeted to promoters through sequence specific DNA binding proteins such as Ume6 (Goldmark et al. 2000; Kent et al. 2001; Shimizu et al. 2003). Second, Isw2-Itc1 acts in conjunction with transcriptional co-repressors such as Rpd3-Sin3 (Fazzio et al. 2001) and Ssn6-Tup1 (Zhang and Reese 2004a, 2004b) to silence genes. Thus deacetylation of lysine residues on the tails of histones that comprise the promoter chromatin is likely to be a feature of genes repressed by Isw2. Thirdly, TBP appears to be a target of Isw2-Itc1-mediated repression (Shimizu et al. 2003; Zhang and Reese 2004a), a function also associated with *Xenopus* (x)-ISWI (Kikyo et al. 2000) and with the Isw1 ATPase in budding yeast (Moreau et al. 2003). Finally, a wide variety of genes, especially those involved in yeast differentiation (meiosis and pseudohyphal growth) are targeted by Isw2-Itc1 (Fazzio et al. 2001; Goldmark et al. 2000; Kent et al. 2001; Ruiz et al. 2003; Sugiyama and Nikawa 2001; Trachtulcova et al. 2000, 2004; Zhang and Reese 2004a, 2004b). The functions associated with ACF-containing complexes can be changed dramatically by the additional subunits, as illustrated for the CHRAC related complexes.

The CHRAC (chromatin accessibility complex) complexes in yeast, flies and mammals vary from ACF only by the addition of two small histone fold proteins. However, this appears to alter the function associated with the ACF from a role in the assembly of repressive chromatin to one that counteracts the spread of repressive heterochromatin and allows movement of replication forks through heterochromatic regions (Bonaldi et al. 2002; Collins et al. 2002; Eberharter et al. 2001; Fyodorov et al. 2004; Kukimoto et al. 2004; McConnell et al. 2004). Moreover, one of the highly conserved histone fold proteins is also a component of the DNA polymerase epsilon complex, specifically targeted to late replicating chromatin (Bonaldi et al. 2002; Eberharter et al. 2001; Fyodorov et al. 2004; Iida and Araki 2004; Kukimoto et al. 2004; McConnell et al. 2004). The relationship between the ISWI complex and DNA polymerase ε is explored in more detail in the next paragraph.

The WSTF (Williams syndrome transcription factor) (Bochar et al. 2000) is also associated with ISWI in the WICH complex and has a related function to CHRAC/ACF in aiding DNA replication through heterochromatin and in antagonizing the spread of heterochromatin (Bozhenok et al. 2002). WSTF is targeted to replication foci throughout S phase. By contrast, ACF1 is not retained at replication sites, suggesting a different targeting mechanism for the two complexes. WSTF interacts directly with PCNA at replication foci, and unlike ACF1, contains a characteristic PCNA binding motif. WSTF then targets ISWI to these regions (Collins et al. 2002; Poot et al. 2004). NoRC functions in a similar way to control replication timing at the heterochromatin rDNA loci (Li et al. 2005; Strohner et al. 2001). Here a transcription termination factor TTF-1 anchors TIP5 to the promoter regions and this brings SNF2H into the complex. TIP5 co-localizes with the basal RNA polymerase I transcription factor UBF in the nucleolus, suggesting regulation of transcription as well as replication timing.

It is not clear what targets the CHRAC-related complexes to heterochromatin. One of the two histone fold proteins in CHRAC complexes is also a component of DNA polymerase epsilon in yeast. An association of the remodelling complex through Pol ε would be an attractive mechanism. However, in yeast a genetic analysis of TPE (telomere position effects on the expression of genes embedded in telomeric heterochromatin) supports opposite functions for Pol ε and yCHRAC (Iida and Araki 2004). Pol ε promotes the stable inheritance of a silent state, while yCHRAC promotes an expressed state. Thus a specific targeting factor remains to be found. However, there may not be one and CHRAC complexes could be targeted directly to specifically modified regions of heterochromatin through the H3 or H4 tails. The link between certain ISWI-containing complexes and DNA replication control is further strengthened by the observation that the TRF2 complex in *Drosophila* is actually required for the expression of PCNA and other factors required for DNA replication (Hochheimer et al. 2002). Thus ISWI complexes control the expression of PCNA and associate directly with PCNA to facilitate replication through heterochromatin.

4.3.2 Yeast Isw1a

In budding yeast, the Isw1a complex, associated with the Ioc3 protein with no known structural motifs, is also associated with repression of gene expression. It associates with yeast promoters and modulates a repressing chromatin structure (Morillon et al. 2003) and may play a role, described for other ISWI complexes, in displacing the TATA binding protein from chromatin (Moreau et al. 2003). The Cbf1 regulatory protein plays an important role in recruiting Isw1 to some promoters (Kent et al. 2004). No other complex described in other eukaryotes appears to be related to Isw1a at the structural level.

4.3.3 NURF and Related Complexes

In addition to activities with a generally repressive function, many species display ISWI-related complexes with defined roles in activating gene expression. These complexes are generally related to NURF in *Drosophila*. Mutations in *nurf301*, encoding the largest subunit associated with ISWI show clear defect in activating a range of genes, including heat shock and homeotic genes (Badenhorst et al. 2002). In addition, mutants in *nurf301* exhibit neoplastic transformation of larval blood cells that causes melanotic tumours to form. A mammalian form of NURF has also been reported (Barak et al. 2003). It differs slightly in that the smallest DmNURF subunit is missing and its effects are limited to nervous tissue, as it is associated with SNF2L. In mouse and humans, SNF2L shows very distinct expression profiles, as in humans it is ubiquitously expressed but limited to neural tissue in mice, although in both species the activity is limited to the nervous system, particularly the brain. This is because humans express a dominant negative splice variant of SNF2L, known as SNF2L+13, which forms an inactive remodelling complex in non-neural tissue (Barak et al. 2004). Mammalian NURF regulates expression of *engrailed*, a homeodomain protein that regulates neuronal development, particularly in the mid-hindbrain. hNURF also promotes neurite outgrowth in culture. Interestingly, NURF 301 is distantly related to the ACF-related family described above, containing a PHD finger, bromodomain, WAC and WAKZ motifs, and at least one unique domain, a HMG-like domain.

4.3.4 Yeast Isw1b

In yeast, the Isw1b complex has a demonstrated role in controlling gene expression, in particular controlling transcription elongation and termination (Morillon et al. 2003). Isw1b is associated with two factors, Ioc2 and Ioc4 (Vary et al. 2003), Ioc2 shares with NURF301 and the ACF-related family a PHD finger. Ioc4 contains a PWWP domain, a member of the royal family of domains associated with methyl lysine binding activity (Maurer-Stroh et al. 2003). In line with this, the Set1 histone methyl transferase activity is required for the association of Isw1b with active genes (Santos-Rosa et al. 2003). It remains to be seen if any of the complexes associated with SNF2H or SNF2L in humans will share the transcription functions associated with yeast Isw1b. SNF2H is also specifically associated with methylated lysine 4 on the histone H3 tails, supporting a related function to Isw1b (Santos-Rosa et al. 2003). An analysis of the contributions of Ioc2 and Ioc4 to Isw1b function shows distinct activities associated with each subunit that together recapitulate Isw1b function (Morillon et al. 2003). Isw1b affects RNA polymerase II function at both the 5′ and 3′ regions of genes. Through recruitment of the Kin28 kinase, that phosphorylates Ser5 on the heptad CTD repeat of RNAPII, Isw1b influences the early elongation phase of transcription and co-transcriptional capping of the nascent transcript. This is influenced by Ioc2 but not Ioc4. In contrast, Isw1b is also required for phosphorylation of Ser2 of the CTD repeat, a mark associated with the 3′ region of genes and co-transcription cleavage and polyadenylation of the transcript, through association of factors such as Rna15 with the CTD. This activity requires Ioc4 but not Ioc2 (Morillon et al. 2003) (Fig. 3).

4.3.5 SNF2h-Cohesin

The final unique complex associated with SNF2H is a cohesin–NuRD complex. SNF2h appears to be anchored to Alu elements via the hRAD21 cohesin protein. The presence of a second ATPase, Mi2 and histone deacetylase activities, within the complex suggests histone modifications accompany cohesin binding. No related activity has yet been described in other organisms. Cohesins associate with newly replicated DNA and are believed to hold sister chromatids together until anaphase. Thus,

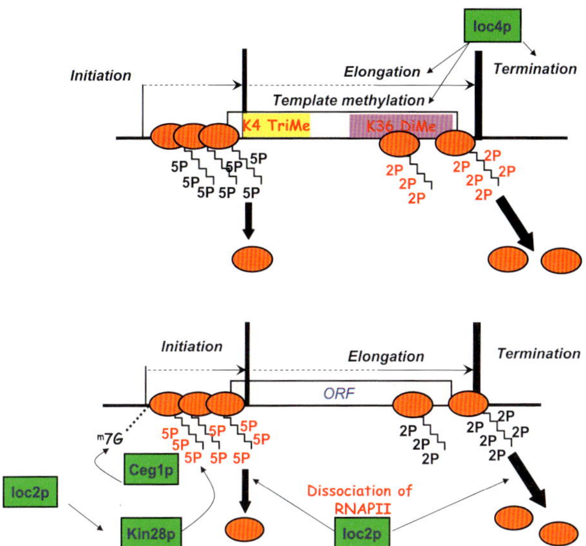

Fig. 3. Schematic showing the effects of the Ioc4 (*top panel*) and Ioc2 (*bottom panel*) components of the yeast (*Saccharomyces cerevisiae*) Isw1b complex on transcription elongation. The open reading frame (ORF) is represented by a *filled rectangle* and the flanking sequences by a *line*. The three phases of transcription, initiation, elongation and termination, are shown separated by *thick black lines*. These represent transitions in the state of phosphorylation (5P or 2P) of the CTD (*zig-zag line*) of RNA polymerase II (RNAPII, *orange oval*) that mark the various phases of transcription. Soon after initiation, RNAPII is phosphorylated at Ser5 by the Kin28 kinase. Ioc2 function is required for this (shown in *red* to indicate that this requires Ioc2). In order to enter productive elongation, RNAPII has to pass a checkpoint (*line*) that ensures that the transcript is capped (^{m7}G) by Ceg1. Ioc2 is envisaged to promote premature termination of transcription at this point unless a positive signal for transcription is received (shown as dissociation of RNAPII). Once RNAPII passes the checkpoint, Ioc4 function in elongation and termination in two ways. Ioc4 promotes Ser2 phosphorylation of the CTD on RNAPII and appropriate methylation of the chromatin template. Together the defects displayed by Ioc2 and Ioc4 are equivalent to those seen in a strain lacking Isw1. Thus the phenotypes associated with *ισω1Δ* are more severe than those in the *ioc2Δ* or the *ioc4Δ* strains

nondisjunction-type defects and/or aneuploidy might also be expected in strains with defective SNF2H/ISWI activity.

4.4 Roles for ISWI Complexes in Development, Differentiation and Disease

Given their central role in chromosome biology, it is not surprising that defects in subunits associated with ISWI proteins give rise to major defects in development or differentiation or are associated with the formation of tumours (Table 2).

4.4.1 CERF

Mammalian SNF2L is unusual in that its activity is limited to the central nervous system and thus most of the defects associated with proteins that form complexes with SNF2L affect the brain and nervous system. The CERF complex is composed of CECR-2 and SNF2L. Mice homozygous for a *cerc2* gene-trap-induced mutation show a high penetrance of the neural tube defect exencephaly, the equivalent of the human condition anencephaly (Banting et al. 2005). CERC contains a DDT (protein interaction domain) and a bromodomain (binds acetylated histone tails), typical of proteins that regulate the activity of chromatin remodelling ATPases. The DDT domain (Doerks et al. 2001) is a signature of virtually all known ISWI-binding partners and is required for Acf1 interaction with ISWI in *Drosophila* (Fyodorov and Kadonaga 2002). CERC2 is known as cat eye syndrome region, candidate 2, one of a cluster of genes duplicated in the human disorder cat eye syndrome. This syndrome is characterized by defects of the eye, heart, anus, kidney, skeleton, face and mental development, although the neural tube develops normally. It is not known if increased gene dosage of *CERC-2* is actually associated with cat eye syndrome.

4.4.2 hNURF

SNF2L is also a component of the hNURF complex and therefore it is not surprising that NURF is enriched in the brain, particularly the hippocampus and cerebellum, and regulates expression of genes that are

Table 2. Roles of ISWI associated factors in disease and development

Organism	Complex	Subunit of complex	Phenotypes	Reference
Drosophila	NURF	NURF301	Gene transcription, expression of heat shock genes and homeotic genes thus associated with developmental defects. Required for male X chromosome structure. Mutants show neoplastic transformation of larval blood leading to melanotic tumours	Badenhorst et al. 2002
Human	WICH	WSTF	Associated with Williams-Beuren syndrome resulting in developmental and mental retardations WSTF is also part of the WINAC complex with BRG1 (SWI/SNF ATPase) and Topoisomerase IIB but not SNF2H. WINAC is not associated with PCNA and the counteraction against the spread of heterochromatin associated with WSTF	Bozhenok et al. 2002; Kitagawa et al. 2003; Poot et al. 2004
Mammals		SNF2H	Snf2h$^{-/-}$ embryos die at the peri-implanation stage. In adults the absence of SNF2h in haematopoietic progenitors prevents cytokine induced erythropoiesis in vitro. Suggests roles in differentiation and development	Stopka and Skoultchi 2003
Mammals	NURF	BPTF	Transcripts for BPTF and SNF2L co-localize to hippocampus and cerebellum in adult mouse brain. SNF2L promotes neurite outgrowth. FAC1 is a truncated version of BPTF associated with Alzheimer's disease.	Barak et al. 2003; Jordan-Sciutto et al. 1999
Mammals	CERF	CERC2	Duplication of *cerc2* region associated with cat eye syndrome. Gene mutation in *cerc2* associated with neural tube defects and anencephaly	Banting et al. 2005

involved in neuronal development in the mid-hindbrain and potentiates neurite outgrowth in culture (Barak et al. 2003). The major component of hNURF, BPTF (bromodomain PHD finger transcription factor) is encoded by *BPTF*, an orthologue of Dm*Nurf301*. *BPTF* on chromosome 17q23, is a known hotspot for chromosomal changes in neuroblastomas and is a prognostic factor for rapid progression of the disease. A truncated form of BPTF, called fetal Alzheimer's clone 1 (FAC) was isolated in a screen for Alzheimer's disease plaques (Bowser et al. 1995). FAC1 levels are up-regulated in motor neurones during development and in the neurodegenerative disease amyotrophic lateral sclerosis (Mu et al. 1997). FAC1 shows sequence-specific DNA binding and may function in transcriptional regulation (Jordan-Sciutto et al. 1999, 2001). Thus PBTF may anchor SNF2L to promoters to regulate and promote neural outgrowth and development.

In summary, differentiation and maturation of neurons during development may involve CERF as well as hNURF. Moreover, NURF in flies and mammals appears to have conserved function. *engrailed* is a target of DmNURF and regulates segment polarity, while the homologues in humans, *engrailed-1*, *engrailed-2*, *wnt-7b* and *pax-2* have a crucial role in cerebellar development (Millen et al. 1995).

4.4.3 DmNURF

Phenotypes associated with flies lacking the large subunit of DmNURF, NURF301 support roles in activation and repression of gene expression and development. NURF301 interacts directly with a number of transcription factors, in particular the GAGA transcription factor (Xiao et al. 2001), suggesting that these interactions anchor ISWI to promoters. NURF301 acts as an enhancer of the bithorax complex *E(bx)* and defects in expression from this sequence element lead to homeotic mutations. Like ISWI, NURF301 is also required for ovary development (Deuring et al. 2000). In males lacking ISWI or NURF301, the male X chromosome is distorted and bloated, implicating NURF in the maintenance of male X chromosome morphology. In flies, X chromosome dosage compensation is achieved by up-regulation of transcription from the male X. This chromosome is uniquely marked by specific acetylation of histone H4 at Lysine 16 that is required for increased transcription and is regulated by the MSL complex (Bhadra et al. 2005) and the MOF histone acetyltransferase (Hilfiker et al. 1997). Paradoxically, H4-K16 acetylation inhibits ISWI activity (Corona et al. 2002), suggesting that NURF does not play a direct role in condensation, although a role in correct nucleosome spacing in higher-order structures is a possible explanation. Although there is no direct evidence of a role for NURF301 in such an activity, in humans hSNF2H and hACF1 regulate chromatin folding into loop domains. The SATB1 transcriptional regulator (special AT-rich sequence binding 1) (Yasui et al. 2002) acts together with the Sin3A/RPD3 deacetylase to target hACF1 and SNF2H to regulate nucleosome positioning over several kilobases. SATB1 defines a class of transcriptional regulators that function as a landing platform for chromatin remodelling enzymes, allowing regulation of large chromatin domains. A defect in this activity could explain the male X chromosome defect. However, as NURF is a known transcription regulatory factor, a more likely function is the transcription of one or more genes encoding regulators of chromatin structure, particularly the higher-order structures that define a chromosome.

NURF and the NURF301 subunit also play a role in tumorigenesis. Flies lacking ISWI or NURF301 exhibit an identical neoplastic transformation of larval blood cells. In the absence of NURF, proliferation and differentiation of haemocytes is triggered, leading to the accumulation of lamellocytes (Badenhorst et al. 2002). This may be related to the transcription-associated functions of NURF, but instead of the gene activation role described above, here NURF appears to repress expression of the genes encoding the signal transduction pathway, specifically the HOP/STAT92E pathway. It remains to be seen whether defect in human ISWI complexes will also be associated with tumorigenesis.

4.4.4 WSTF and WCRF

Williams syndrome is a rare autosomal dominant hereditary disorder with multiple symptoms, including typical congenital vascular lesions, elfin face, mental retardation and growth deficiency (Lu et al. 1998). The transient appearance of infantile aberrant vitamin D metabolism and hypercalcaemia are also documented. The syndrome is associated with deletions at chromosome 7q11.23 and there are several candidate genes within this region. One gene, *WSTF*, is the most likely candidate that when defective leads to the diverse phenotypes that make up the syndrome. Currently, WSTF is a component of a least two chromatin remodelling complexes, WINAC and WICH. These complexes are distinguished by the nature of the ATPase: WINAC contains a SWI/SNF type enzyme (Lu et al. 1998) while WICH contains SNF2H (ISWI) (Bozhenok et al. 2002; Poot et al. 2004). Both complexes contain, or are associated with, DNA replication-related factors. WINAC and WICH are both required for normal progression through S phase. WINAC has an additional role in gene expression, being required for the recruitment of the vitamin D receptor to vitamin D-regulated promoters (Kitagawa et al. 2003), explaining one aspect of the syndrome. An interesting question is whether WICH, or the related WCRF complex containing a WSTF-related factor will also regulate the expression of specific genes.

4.5 ISWI Complexes and the Histone Code

Unlike the SWI/SNF chromatin remodelling ATPase, the histone tails are essential for ISWI enzymes to mobilize nucleosomes (Hamiche et al. 2001). In addition, covalent modifications to the tails influence ISWI remodelling (Clapier et al. 2001, 2002) and a role for ISWI in presenting histone tails for covalent modification has been proposed (Grune et al. 2003). Modifications to histone H4 tail, particularly lysine acetylation, tend to reduce the ability of ISWI to remodel nucleosomes (Clapier et al. 2001, 2002; Corona et al. 2002). By contrast, methylation of lysines on histone H3 tend to promote the association of ISWI with chromatin (Santos-Rosa et al. 2003). Thus ISWI complexes are intimately linked to the histone code. Again, this is best understood for the yeast Isw1b enzyme and this final section will discuss this aspect of ISWI biology in more detail.

The ISWI proteins are unusual in that they contain no well-defined element that would interact with modified residues on histone tails, such as bromodomains of the SWI/SNF family that bind to acetylated lysines (Dhalluin et al. 1999) or the chromodomains of the Mi-2/CHD family that bind to methylated lysines (Pray-Grant et al. 2005). This role has been proposed for the SANT domain but no specific interactions have yet been characterized. Nevertheless, there is good evidence that some ISWI complexes, for instance Isw1b in yeast and hSNF2H, show specific interactions with the histone H3 tail only when lysine 4 is methylated (Santos-Rosa et al. 2003). Lysine 4 methylation is a mark of active genes (Bernstein et al. 2002; Santos-Rosa et al. 2002), supporting functions in transcription elongation for Isw1b (Morillon et al. 2003) and for one or more of the complexes containing SNF2H. However, it is not known whether this is a function the ISWI protein itself or an interacting protein within the complex. For instance, Ioc4 within Isw1b contains a PWWP domain, a putative methyl lysine-binding protein (Maurer-Stroh et al. 2003). Other ISWI-associated proteins contain bromodomains (e.g. BPTF in hNURF, CERC-2 in CERF, ACF1 in ACF1 and CHRAC complexes, WRCF and WSTF in WICH) or chromodomains (e.g. Mi2 in SNF2H-cohesin-NuRD complex), suggesting interactions with modified histone tails.

In addition to a role for lysine 4 methylation in recruitment of Isw1b to chromatin, the ATPase activity of Isw1 is also required for some patterns of histone lysine methylation, for instance, H3K4me2, H3K4me3 and H3K36me2, all marks on active genes (Morillon et al. 2003). In this respect, the subunits of Isw1b do not behave equivalently. Ioc4 is required for H3K4 and H3K36 methylation. By contrast, Ioc2 influences the distribution of these marks over the coding region of active genes. For instance, marks such as K36me2 that accumulate at the 3′ region of genes and are associated with the termination of transcription tend to accumulate at the 5′ region of genes in *ioc2Δ* mutants. These methylation marks are proposed to recruit factors that influence the various stages of transcription elongation and termination. Thus the interplay between Ioc2 and Ioc4 in association with Isw1 ATPase controls elongating RNA polymerase II, sequentially linking each stage of the transcription cycle, linking events at the 5′ and 3′ regions of genes and controlling termination of transcription.

Acknowledgements. The author thanks The Wellcome Trust, CRUK and the BBSRC for research funding.

References

Badenhorst P, Voas M, Rebay I, Wu C (2002) Biological functions of the ISWI chromatin remodeling complex NURF. Genes Dev 16:3186–3198

Banting GS, Barak O, Ames TM, Burnham AC, Kardel MD, Cooch NS, Davidson CE, Godbout R, McDermid HE, Shiekhattar R (2005) CECR2, a protein involved in neurulation, forms a novel chromatin remodeling complex with SNF2L Hum Mol Genet 14:513–524

Barak O, Lazzaro MA, Lane WS, Speicher DW, Picketts DJ, Shiekhattar R (2003) Isolation of human NURF: a regulator of Engrailed gene expression. EMBO J 22:6089–100

Barak O, Lazzaro MA, Cooch NS, Picketts DJ, Shiekhattar R (2004) A tissue-specific naturally occurring human SNF2L variant inactivates chromatin remodeling. J Biol Chem 279:45130–45138

Becker PB, Horz W (2002) ATP-dependent nucleosome remodelling. Annu Rev Biochem 71:247–273

Bernstein BE, Humphrey EL, Erlich RL, Schneider R, Bouman P, Liu JS, Kouzarides T, Schreiber SL (2002) Methylation of histone H3 Lys 4 in coding regions of active genes. Proc Natl Acad Sci U S A 99:8695–700

Bhadra MP, Bhadra U, Kundu J, Birchler JA (2005) Gene expression analysis of the function of the male-specific lethal complex in Drosophila. Genetics 169:2061–2074

Bochar DA, Savard J, Wang W, Lafleur DW, Moore P, Cote J, Shiekhattar R (2000) A family of chromatin remodeling factors related to Williams syndrome transcription factor. PNAS 97:1038–1043

Bonaldi T, Langst G, Strohner R, Becker PB, Bianchi ME (2002) The DNA chaperone HMGB1 facilitates ACF/CHRAC-dependent nucleosome sliding. EMBO J 21:6865–6873

Bowser R, Giambrone A, Davies P (1995) FAC1, a novel gene identified with the monoclonal antibody Alz50, is developmentally regulated in human brain. Dev Neurosci 17:20–37

Boyer LA, Langer MR, Crowley KA, Tan S, Denu JM, Peterson CL (2002) Essential role for the SANT domain in the functioning of multiple chromatin remodelling enzymes. Mol Cell 10:935–942

Bozhenok L, Wade PA, Varga-Weisz P (2002) WSTF-ISWI chromatin remodeling complex targets heterochromatic replication foci. EMBO J 21:2231–2241

Clapier CR, Langst G, Corona DF, Becker PB, Nightingale KP (2001) Critical role for the histone H4N terminus in nucleosome remodeling by ISWI. Mol Cell 3:239–245

Clapier CR, Nightingale KP, Becker PB (2002) A critical epitope for substrate recognition by the nucleosome remodeling ATPase ISWI. Nucleic Acids Res 30:649–655

Collins N, Poot R, Kukimoto I, Garcia-Jimenez C, Dellaire G, Varga-Weisz P (2002) An ACF1-ISWI chromatin-remodeling complex is required for DNA replication through heterochromatin. Nature Genetics 32:627–632

Corona DF, Clapier CR, Becker PB, Tamkun JW (2002) Modulation of ISWI function by site-specific histone acetylation. EMBO Rep 3:242–247

Cuperus G, Shore D (2002) Restoration of silencing in Saccharomyces cerevisiae by tethering of a novel Sir2-interacting protein Esc8. Genetics 162:633–645

De la Cruz X, Lois S, Sanchez-Molina S, Martinez-Balbas MA (2005) Do protein motifs read the histone code? Bioessays 27:164–175

Deuring R, Fanti L, Armstrong JA, Sarte M, Papoulas O, Prestel M, Daubresse G, Verardo M, Moseley SL, Berlocol M et al. (2000) The ISWI chromatin-remodeling protein is required for gene expression and the maintenance of higher order chromatin structure in vivo. Mol Cell 5:355–365

Dhalluin C, Carlson JE, Zeng L, He C, Aggarwal AK, Zhou M-M, Zhou M-M (1999) Structure and ligand of a histone acetyltransferase bromodomain. Nature 399:491–496

Doerks T, Copley R, Bork P (2001) DDT – a novel domain in different transcription and chromosome remodeling factors. Trends Biochem Sci 26:145–146

Eberharter A, Ferrari S, Langst G, Straub T, Imhof A, Varger-Weisz PD, Wilm M, Becker PB (2001) Acf-1, the largest subunit of CHRAC, regulates ISWI-induced nucleosome remodeling. EMBO J 20:3781–3784

Eberharter A, Vetter I, Ferreira R, Becker PB (2004) ACF1 improves the effectiveness of nucleosome mobilization by ISWI through PHD-histone contacts. EMBO J 23:4029–4039

Fazzio TG, Kooperberg C, Goldmark JP, Neal C, Basom R, Delrow J, Tsukiyama T (2001) Widespread collaboration of Isw2 Sin3-Rpd3 chromatin remodeling complexes in transcriptional repression. Mol Cell Biol 21:6450–6460

Fyodorov DV, Kadonaga JT (2002) Binding of Acf1 to DNA Involves a WAC motif is important for ACF-mediated chromatin assembly. Mol Cell Biol 22:6344–6353

Fyodorov DV, Blower MD, Karpen GH, Kadonaga JT (2004) Acf1 confers unique activities to ACF/CHRAC and promotes the formation rather than disruption of chromatin in vivo. Genes Dev 18:170–183

Gavin AC, Bosche M, Krause R et al. (2002) Functional organization of the yeast proteome by systematic analysis of protein complexes. Nature 415:141–147

Goldmark JP, Fazzio TG, Estep PW, Church GM, Tsukiyama T (2000) The Isw2 chromatin remodeling complex represses early meiotic genes upon recruitment of Ume6p. Cell 103:423–433

Grune T, Brzeski J, Eberharter A, Clapier CR, Corona DFV, Becker PB, Muller CW (2003) Crystal structure functional analysis of a nucleosome recognition module of the remodeling factor ISWI. Mol Cell 12:449–460

Hakimi M-A, Bochar DA, Schmiesing JA, Dong Y, Barak OG, Speicher DW, Yokomori K, Shiekhattar R (2002) A chromatin remodelling complex that loads cohesin onto human chromosomes. Nature 418:994–998

Hamiche A, Sandaltzopoulos R, Gdula DA, Wu K (1999) ATP-dependent histone octamer sliding mediated by the chromatin remodeling complex NURF. Cell 97:833–842

Hamiche A, Kang J-G, Dennis C, Xiao H, Wu C (2001) Histone tails modulate nucleosome mobility and regulate ATP-dependent nucleosome sliding by NURF. PNAS 98:14316–14321

Hilfiker A, Hilfiker Kleiner D, Pannuti A, Lucchesi JC (1997) mof, a putative acetyl transferase gene related to the Tip60 MOZ human genes and to the SAS genes of yeast, is required for dosage compensation in Drosophila. EMBO J 16:2054–2060

Hochheimer A, Zhou S, Zheng S, Holmes MC, Tjian R (2002) TRF2 associates with DREF and directs promoter-selective gene expression in Drosophila. Nature 420:439–445

Iida T, Araki H (2004) Noncompetitive counteractions of DNA polymerase epsilon ISW2/yCHRAC for epigenetic inheritance of telomere position effect in Saccharomyces cerevisiae. Mol Cell Biol 24:217–227

Jordan-Sciutto KL, Dragich JM, Rhodes JL, Bowser R (1999) Fetal Alz-50 Clone 1, a novel zinc finger protein binds a specific DNA sequence acts as a transcriptional regulator. J Biol Chem 274:35262–35268

Jordan-Sciutto K, Rhodes J, Bowser R (2001) Altered subcellular distribution of transcriptional regulators in response to Abeta peptide and during Alzheimer's disease. Mech Ageing Dev 123:11–20

Kent NA, Karabetsou N, Politis PK, Mellor J (2001) In vivo chromatin remodeling by yeast ISWI homologs Isw1p and Isw2p. Genes Dev 15:619–626

Kent NA, Eibert SM, Mellor J (2004) Cbf1p is required for chromatin remodeling at promoter-proximal CACGTG motifs in yeast. J Biol Chem 279:27116–27123

Kikyo N, Wade PA, Guschin D, Ge H, Wolffe AP (2000) Active remodeling of somatic nuclei in egg cytoplasm by the nucleosomal ATPase ISWI. Science 289:2360–2362

Kitagawa H, Fujiki R, Yoshimura K, Mezaki Y, Uematsu Y, Matsui D, Ogawa S, Unno K, Okubo M, Tokita A (2003) The chromatin-remodeling complex

WINAC targets a nuclear receptor to promoters is impaired in Williams syndrome. Cell 113:905–917
Kukimoto I, Elderkin S, Grimaldi M, Oelgeschlager T, Varga-Weisz PD (2004) The histone-fold protein complex CHRAC-15/17 enhances nucleosome sliding assembly mediated by ACF. Mol Cell 13:265–277
LeRoy G, Orphanides G, Lane WS, Reinberg D (1998) Requirement of RSF FACT for transcription of chromatin templates in vitro. Science 282:1900–1904
LeRoy G, Loyola A, Lane WS, Reinberg D (2000) Purification characterization of a human factor that assembles remodels chromatin. J Biol Chem 275:14787–14790
Li J, Santoro R, Koberna K, Grummt I (2005) The chromatin remodeling complex NoRC controls replication timing of rRNA genes. EMBO J 24:120–127
Loyola A, Huang J-Y, LeRoy G, Hu S, Wang Y-H, Donnelly RJ, Lane WS, Lee S-C, Reinberg D (2003) Functional analysis of the subunits of the chromatin assembly factor RSF. Mol Cell Biol 23:6759–6768
Lu X, Meng X, Morris CA, Keating MT (1998) A novel human gene, WSTF, is deleted in Williams syndrome. Genomics 54:241–249
Lusser A, Urwin DL, Kadonaga JT (2005) Distinct activities of CHD1 and ACF in ATP-dependent chromatin assembly 12:160–166
MacCallum DE, Losada A, Kobayashi R, Hirano T (2002) ISWI remodeling complexes in Xenopus egg extracts: identification as major chromosomal components that are regulated by INCENP-aurora B. Mol Biol Cell 13:25–39
Martinez Balbas MA, Tsukiyama T, Gdula D, Wu C (1998) Drosophila NURF-55, a WD repeat protein involved in histone metabolism. Proc Natl Acad Sci U S A 95:132–137
Maurer-Stroh S, Dickens N, Hughes-Davies L, Kouzarides T, Eisenhaber F, Ponting C (2003) The tudor domain 'royal family': tudor, plant agent chromo PWWP MBT domains. Trends Bioch Sci 28:69–74
McConnell AD, Gelbart ME, Tsukiyama T (2004) Histone fold protein Dls1p is required for Isw2-dependent chromatin remodeling in vivo. Mol Cell Biol 24:2605–2613
Mellor J, Morillon A (2004) ISWI complexes in yeast. Bioch Biophys Acta 1677:100–112
Millen K, Hui C, Joyner A (1995) A role for En-2 and other murine homologues of Drosophila segment polarity genes in regulating positional information in the developing cerebellum. Development 121:3935–3945
Moreau J-L, Lee M, Mahachi N, Vary JC Jr, Mellor J, Tsukiyama T, Goding C (2003) Regulated displacement of TBP from the PHO8 promoter in vivo requires Cbf1 and the Isw1 chromatin remodeling complex. Mol Cell 11:1609–1620

Morillon A, Karabetsou N, O'Sullivan J, Kent NA, Proudfoot NJ, Mellor J (2003) Isw1 chromatin remodeling ATPase coordinates transcription elongation and termination by RNA polymerase II. Cell 115:425–435

Mu X, Springer JE, Bowser R (1997) FAC1 Expression localization in motor neurons of developing adult, and amyotrophic lateral sclerosis spinal cord. Exp Neurol 146:17–24

Poot RA, Bozhenok L, van den Berg DLC, Steffensen S, Ferreira F, Grimaldi M, Gilbert N, Ferreira J, Varga-Weisz PD (2004) The Williams syndrome transcription factor interacts with PCNA to target chromatin remodelling by ISWI to replication foci. Nat Cell Biol 6:1236–1244

Pray-Grant MG, Daniel JA, Schieltz D, Yates JR III, Grant PA (2005) Chd1 chromodomain links histone H3 methylation with SAGA-SLIK-dependent acetylation. Nature 433:434–438

Ruiz C, Escribano V, Morgado E, Molina M, Mazon MJ (2003) Cell-type-dependent repression of yeast a-specific genes requires Itc1p, a subunit of the Isw2p-Itc1p chromatin remodelling complex. Microbiology 149:341–351

Santos-Rosa H, Schneider R, Bannister AJ, Sherriff J, Bernstein BE, Emre NC, Schreiber SL, Mellor J, Kouzarides T (2002) Active genes are tri-methylated at K4 of histone H3. Nature 419:407–411

Santos-Rosa H, Schneider R, Bernstein BE, Karabetsou N, Morillon A, Weise C, Schreiber SL, Mellor J, Kouzarides T (2003) Methylation of histone H3K4 mediates association of the Isw1p ATPase with chromatin. Mol Cell 12:1325–1332

Schwanbeck R, Xiao H, Wu C (2004) Spatial contacts nucleosome step movements induced by the NURF chromatin remodeling complex. J Biol Chem 279:39933–39941

Shimizu M, Takahashi K, Lamb TM, Shindo H, Mitchell AP (2003) Yeast Ume6p repressor permits activator binding but restricts TBP binding at the HOP1 promoter. Nucleic Acids Res 31:3033–3037

Stopka T, Skoultchi AI (2003) The ISWI ATPase Snf2h is required for early mouse development. PNAS 100:14097–14102

Strohner R, Nemeth A, Jansa P, Hofmann-Rohrer U, Santoro R, Langst G, Grummt I (2001) NoRC – a novel member of mammalian ISWI-containing chromatin remodeling machines. EMBO J 20:4892–4900

Sugiyama M, Nikawa J (2001) The Saccharomyces cerevisiae Isw2p-Itc1p complex represses INO1 expression and maintains cell morphology. J Bacteriol 183:4985–4993

Trachtulkova P, Janatova I, Kohlwein SD, Hasek J (2000) Saccharomyces cerevisiae gene ISW2 encodes a mircotubule-interacting protein required for pre-meiotic DNA replication. Yeast 16:35–47

Trachtulcova P, Frydlova I, Janatova I, Hasek J (2004) The absence of the Isw2p-Itc1p chromatin-remodelling complex induces mating type-specific Flo11p-independent invasive growth of Saccharomyces cerevisiae. Yeast 21:389–401

Tsukiyama T, Palmer J, Landel CC, Shiloach J, Wu C (1999) Characterization of the imitation switch subfamily of ATP-dependent chromatin-remodeling factors in Saccharomyces cerevisiae. Genes Dev 13:686–697

Vary JC Jr, Gangaraju VK, Qin J, Landel CC, Kooperberg C, Bartholomew B, Tsukiyama T (2003) Yeast Isw1p forms two separable complexes in vivo. Mol Cell Biol 23:80–91

Xiao H, Sandaltzopoulos R, Wang HM, Hamiche A, Ranallo R, Lee KM, Fu D, Wu C (2001) Dual function of largest NURF subunit NURF 301 in nucleosome sliding and transcriptional interactions. Mol Cell 8:531–543

Yasui D, Miyano M, Cai S, Varga-Weisz P, Kohwi-Shigematsu T (2002) SATB1 targets chromatin remodelling to regulate genes over long distances. Nature 419:641–645

Zhang Z, Reese JC (2004a) Redundant mechanisms are used by Ssn6-Tup1 in repressing chromosomal gene transcription in Saccharomyces cerevisiae. J Biol Chem 279:39240–39250

Zhang Z, Reese JC (2004b) Ssn6–Tup1 requires the ISW2 complex to position nucleosomes in Saccharomyces cerevisiae. EMBO J 23:2246–2257

5 How Is Epigenetic Information on Chromatin Inherited After DNA Replication?

Y. Nakatani, H. Tagami, E. Shestakova

Abstract. Although most somatic cells have identical genetic information, gene expression profiles are quite distinct in each cell type. The gene expression profiles are considered to be determined mainly by chromatin-encoded epigenetic information that includes histone modifications, histone variants, and factors such as HP1 and polycomb group proteins that organize higher-ordered chromatin structures. To gain insights into how such epigenetic information on chromatin is inherited on daughter DNA strands after DNA replication, we have purified the preassembled form of histone H3 by immunoaffinity purification. The histone H3 complex contains the two histone H3-H4 chaperones CAF1 and ASF1. Surprisingly, the H3 complex also contains a pair of H3-H4 dimers. This observation is striking because histones H3-H4 are known to exist as tetramers in solution. Since histones H3-H4 in the predeposition complex exist as a dimer, this raises the possibility that the H3-H4 dimer in the complex pairs with a parental H3-H4 dimer, assembling the de novo-synthesized and parental H3-H4 dimers in the same nucleosome. Based on these results, we propose a semi-conservative model of nucleosome duplication, which allows for segregation of parental H3-H4 dimers with encoded epigenetic information evenly to daughter DNA strands.

5.1 Epigenetic Information on Chromatin

While most somatic cells have identical genetic information (DNA sequences), gene expression patterns are quite distinct depending on the cell types. The differences in gene expression patterns are mainly determined by chromatin structure. Chromatin is characterized as heterochromatin or euchromatin based on observed cytological condensation of chromatin (Wolffe 1998). While most heterochromatic regions appear to be transcriptionally inactive, euchromatic regions (noncondensed regions) contain both transcriptionally active and inactive chromatin. Thus, there is no direct correlation between chromatin condensation and transcription. In this manuscript, we use the terminology "transcriptionally active and inactive chromatin" to distinguish the chromatin regions by their functions.

Transcriptionally active and inactive regions of chromatin are thought to be determined by specific modifications of histones, such as acetylation, phosphorylation, and methylation (Jenuwein and Allis 2001; Grewal and Elgin 2002; Turner 2002; Felsenfeld and Groudine 2003). For instance, histone H3 methylated at K9 is predominantly found in the inactive chromatin, whereas histone H3 acetylated at K14 is found in the active chromatin. In addition to the modification of histones, histone variants are also important determinants of specialized chromatin structures. For instance, macroH2A is predominantly found in transcriptionally inactive chromatin, while H3.3 and H2AZ is deposited in active chromatin (Kamakaka and Biggins 2005). Histone modifications and variants are thought to encode epigenetic information, which contributes to the determination of cell types. A central question is how epigenetic information on chromatin is maintained after DNA replication. When DNA is replicated, an equal amount of parental histones and de novo synthesized histones are deposited onto daughter DNA strands. So the question becomes how is epigenetic information inherited by daughter chromatin?

5.2 Nonconservative Nucleosome Segregation Model

According to the popular model, newly replicated DNA is assembled into chromatin using parental histones or de novo synthesized histones in separate pathways (reviewed in Wolffe 1998). The first pathway involves transfer of the parental $(H3–H4)_2$ tetramers onto one of the daughter DNA strands. Each parental nucleosome is postulated to be transiently disrupted into a $(H3–H4)_2$ tetramer and two H2A–H2B dimers during passage of the replication fork (Gruss et al. 1993). Subsequently, each parental $(H3–H4)_2$ tetramer is assembled randomly onto either of the daughter DNA strands (Cusick et al. 1984; Sogo et al. 1986; Burhans et al. 1991). The second pathway involves deposition of newly synthesized $(H3–H4)_2$ tetramers on daughter DNA strands (Kaufman and Almouzni 2000; Verreault 2000; Krude and Keller 2001). Thus, by this model, parental histones and de novo-synthesized histones are deposited into distinct nucleosomes (Fig. 1A). If this is the case, parental histones with epigenetic information would not be segregated evenly to daughter DNA strands.

5.3 Semiconservative Nucleosome Segregation Model

Since histones H3–H4 exist as tetramers in solution, it was long thought that de novo synthesized histones also exist as tetramers. However, very surprisingly, we discovered that pre-deposit form of histones H3–H4 exist as dimers. We have realized this important issue in a course of characterization of predeposition histone complexes. For purification, the FLAG-HA-epitope sequence was fused to the C-terminal end of histone H3 (e-H3) (Fig. 2A), and the fusion protein was stably expressed in HeLa cells. Importantly, e-H3.1 co-localizes with mitotic with chromosomes (Fig. 2B). This indicates that the epitope tag attached to histone H3 does not interfere with nucleosome assembly in vivo, and thus, purified complexes could be relevant. To purify preassembled forms of histone H3, we used nuclear extracts, which do not contain chromatin, as starting materials. e-H3 was purified from nuclear extracts by two-step immunoaffinity purification (FLAG-antibody immunoprecipitation followed by HA-antibody immunoprecipitation) (Nakatani and Ogryzko 2003). Mass spectrometric analyses revealed that the H3.1 complex con-

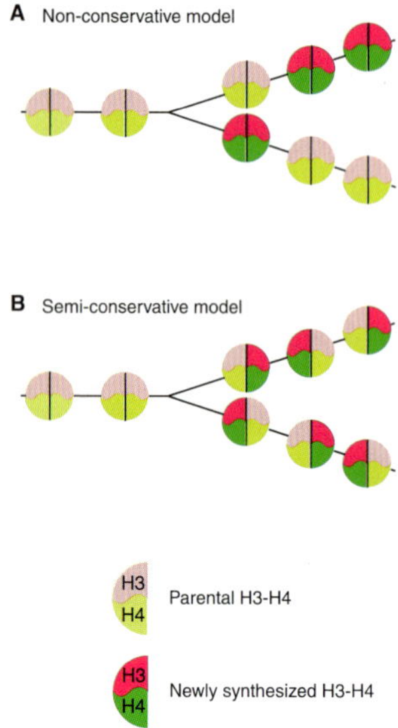

Fig. 1A,B. Models for chromatin duplication. **A** The nonconservative segregation model. **B** The semiconservative model

tains two distinct histone chaperones for H3–H4; all subunits of CAF-1 (p150, p60, and p48) (Smith and Stillman 1989) and two isoforms of ASF1 (ASF1A and B) (Munakata et al. 2000; Sillje and Nigg 2001; Tyler et al. 2001; Mello et al. 2002). Moreover, the complex contains NASP, which is a histone H1-binding protein (Richardson et al. 2000; Alekseev et al. 2003). The complex also contains HAT1, which acetylates predeposited histone H3–H4. Newly synthesized histones are known to be acetylated by HAT1 (Verreault 2000). Although the role of acetylation by HAT1 is unclear, the acetylation is considered to be linked to nucleosome assembly, rather than transcriptional control. We have demonstrated that histones H3 and H4 in this complex exist as dimers in the following ways. As illustrated in Fig. 3A, if histone H3–H4 existed as a tetramer, the complex would have both epitope-tagged and untagged H3 because the epitope-tagged histones occupy only the small fractions compared with the endogenous histones in our cell lines. In-

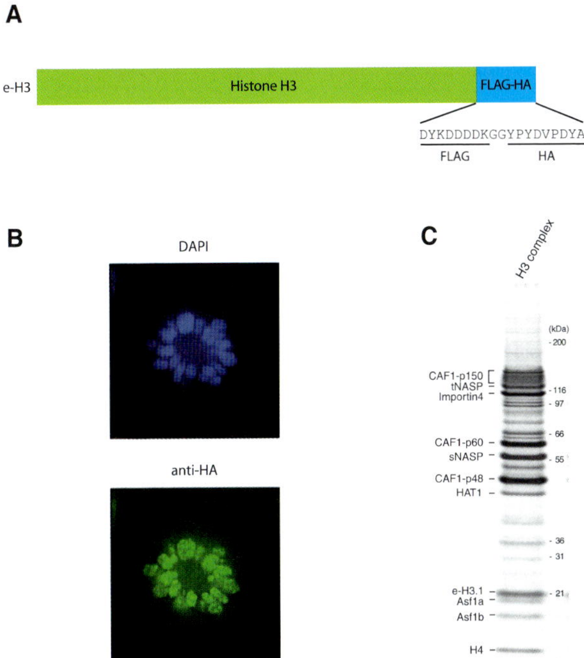

Fig. 2A–C. **A** Construction of FLAG-HA-tagged histone H3 (e-H3). **B** e-H3 co-localizations with mitotic chromosomes. HeLa cells expressing e-H3 are stained with DAPI (*top*) and anti-HA antibody (*bottom*). **C** Silver staining of the H3 complex. (Adapted from Tagami et al. 2004)

deed, mononucleosome core particles purified by anti-FLAG antibody immunoprecipitation contain both epitope-tagged and untagged H3 by Western blotting with anti-H3 antibody (Fig. 3B, lane 1). In contrast, the H3 complex contains only epitope-tagged H3 (lane 2). From this we conclude that histones H3 and H4 in the H3 complex are dimers, rather than tetramers.

These findings raise the possibility that parental $(H3–H4)_2$ tetramers are disrupted and segregated evenly onto daughter DNA strands as H3–

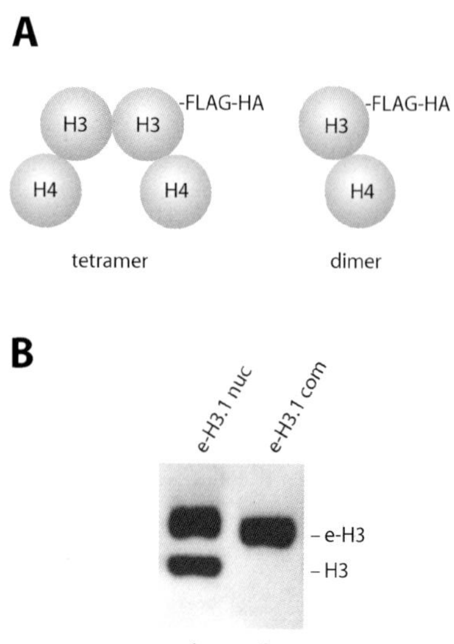

Fig. 3A,B. Since untagged H3 is far more abundant than epitope-tagged H3 in the cell lines employed, the epitope-tagged H3 could preferentially pair with untagged H3 to form the H3–H4 tetramer. Thus, mononucleosomes contain epitope-tagged and untagged H3 (*lane 1*). In contrast, the H3 complex has no detectable untagged H3, indicating that histones H3 and H4 in these complexes exist as a dimer. (Adapted from Tagami et al. 2004)

H4 dimers, which are then paired with de novo-synthesized H3–H4 dimers by the action of the H3 complex (Fig. 1B). This model allows deposition of mixed parental and de novo synthesized histones in the same nucleosome particle, ensuring inheritance of parental H3–H4 dimers in all nucleosome core particles formed on newly synthesized DNA strands. Although it might be too early to discuss this issue, if the parental H3–H4 dimer is deposited with the de novo synthesized H3–H4 dimer in the same nucleosome, an important question becomes how histone modifications on parental H3–H4 would be copied onto the de novo synthesized H3–H4 in the same nucleosome. In the case of DNA methylation, methylated DNA becomes hemi-methylated form after replication. Thereafter, hemi-methylated DNA would be rapidly converted to the homo-methylated form because DNA methyltransferases have higher affinity for hemi-methylated DNAs. Likewise, there might be histone modifying enzymes that preferentially recognize newly synthesized nucleosomes, which consist of the parental and de novo synthesized H3–H4 dimers, allowing the modification on the parental H3–H4 to be copied on the de novo synthesized H3–H4. A mechanism such as this would be required to maintain epigenetic information on chromatin.

Acknowledgements. Y.N. is in part supported by grants from NIH (GM065939-02).

References

Alekseev OM, Bencic DC, Richardson RT, Widgren EE, O'Rand MG (2003) Overexpression of the Linker histone-binding protein tNASP affects progression through the cell cycle. J Biol Chem 278:8846–8852

Burhans WC, Vassilev LT, Wu J, Sogo JM, Nallaseth FS, DePamphilis ML (1991) Emetine allows identification of origins of mammalian DNA replication by imbalanced DNA synthesis, not through conservative nucleosome segregation. EMBO J 10:4351–4360

Cusick ME, DePamphilis ML, Wassarman PM (1984) Dispersive segregation of nucleosomes during replication of simian virus 40 chromosomes. J Mol Biol 178:249–271

Felsenfeld G, Groudine M (2003) Controlling the double helix. Nature 421:448–453

Grewal SI, Elgin SC (2002) Heterochromatin: new possibilities for the inheritance of structure. Curr Opin Genet Dev 12:178–187

Gruss C, Wu J, Koller T, Sogo JM (1993) Disruption of the nucleosomes at the replication fork. EMBO J 12:4533–4545

Jenuwein T, Allis CD (2001) Translating the histone code. Science 293:1074–1080

Kamakaka RT, Biggins S (2005) Histone variants: deviants? Genes Dev 19:295–310

Kaufman PD, Almouzni G (2000) DNA replication, nucleotide excision repair and nucleosome assembly. In: Elgin SCR, Workman JL (eds) Chromatin structure and gene expression. Oxford University Press, Oxford, pp 24–48

Krude T, Keller C (2001) Chromatin assembly during S phase: contributions from histone deposition DNA replication and the cell division cycle. Cell Mol Life Sci 58:665–672

Mello JA, Sillje HH, Roche DM, Kirschner DB, Nigg EA, Almouzni G (2002) Human Asf1 and CAF-1 interact and synergize in a repair-coupled nucleosome assembly pathway. EMBO Rep 3:329–334

Munakata T, Adachi N, Yokoyama N, Kuzuhara T, Horikoshi M (2000) A human homologue of yeast anti-silencing factor has histone chaperone activity. Genes Cells 5:221–233

Nakatani Y, Ogryzko V (2003) Immunoaffinity purification of mammalian protein complexes. Methods Enzymol 370:430–444

Richardson RT, Batova IN, Widgren EE, Zheng LX, Whitfield M, Marzluff WF, O'Rand MG (2000) Characterization of the histone H1-binding protein NASP, as a cell cycle-regulated somatic protein. J Biol Chem 275:30378–30386

Sillje HH, Nigg EA (2001) Identification of human Asf1 chromatin assembly factors as substrates of Tousled-like kinases. Curr Biol 11:1068–1073

Smith S, Stillman B (1989) Purification and characterization of CAF-I, a human cell factor required for chromatin assembly during DNA replication in vitro. Cell 58:15–25

Sogo JM, Stahl H, Koller T, Knippers R (1986) Structure of replicating simian virus 40 minichromosomes. The replication fork, core histone segregation and terminal structures. J Mol Biol 189:189–204

Tagami H, Ray-Gallet D, Almouzni G, Nakatani Y (2004) Histone H3.1 and H3.3 complexes mediate nucleosome assembly pathways dependent or independent of DNA synthesis. Cell 116:51–61

Turner BM (2002) Cellular memory and the histone code. Cell 111:285–291

Tyler JK, Collins KA, Prasad-Sinha J, Amiott E, Bulger M, Harte PJ, Kobayashi R, Kadonaga JT (2001) Interaction between the Drosophila CAF-1 and ASF1 chromatin assembly factors. Mol Cell Biol 21:6574–6584

Verreault A (2000) De novo nucleosome assembly: new pieces in an old puzzle. Genes Dev 14:1430–1438

Wolffe A (1998) Chromatin: structure and function, 3rd edn. Academic Press, London

6 Polycomb Silencing Mechanisms and Genomic Programming

V. Pirrotta

Abstract. Polycomb complexes, best known for their role in the epigenetic silencing of homeotic genes, are now known to regulate a large number of functions in organisms from flies to man. They control transcription activators, pattern-forming genes, maintenance of stem cells and are implicated in cell proliferation and oncogenesis. Our understanding of Polycomb mechanisms derives principally from the study of homeotic genes in *Drosophila*, where they act in an all-or-none fashion to silence expression in inappropriate parts of the organism. This review summarizes what has been learned from homeotic genes and examines the possible extensions of Polycomb mechanisms to allow for dynamic regulatory behavior and the reprogramming of silenced chromatin states.

6.1 Introduction

Developmental processes offer many examples of signals that induce a permanent change in the developmental potential of cells and their cellular progeny. Early circumstances thus alter the competence of certain cells that are otherwise not distinguishable from their neighbors. Polycomb mechanisms are involved in some of the better understood examples of this kind of genomic programming.

Polycomb silencing mechanisms were first discovered in *Drosophila* as functions essential for the regulated expression of homeotic genes during development. We now know that the Polycomb Group (PcG) proteins and their complexes are essentially conserved from insects to mammals. We also know that, in flies as well as in vertebrates, many genes in addition to the homeotics are targets of PcG mechanisms. Current analysis by chromatin immunoprecipitation combined with genomic tiling arrays (ChIP-on-chip) reveals that, in *Drosophila*, the majority of these target genes encode proteins that regulate gene expression but others encode growth factors, signaling components, morphogenetic regulators such as TGF-β, Wingless, Epidermal Growth Factor Receptor, Hedgehog, Engrailed, and some encode components of the Polycomb complexes themselves (Y. Schwartz, T. Kahn and V. Pirrotta, unpublished results). It is likely that what PcG proteins do at some of these loci will turn out to be surprisingly different from what they do at the classical homeotic loci. Work in many laboratories in the past 15 years has made considerable progress in understanding the molecular bases for PcG regulation of homeotic loci and, while complete consensus has not been reached on the finer details, the basic features of the mechanism are well established for these loci. The following account is a brief synopsis of what I will call the homeotic paradigm.

6.2 The Homeotic Paradigm

In *Drosophila*, genetic and developmental studies have shown that each homeotic gene must be expressed only in its appropriate segmental domain and be repressed in other segments of the embryo. This segmental pattern is established in the first few hours after fertilization and is maintained in later development. The prototypical homeotic gene, *Ultrabithorax* (*Ubx*), is activated by segmentation gene products and repressed by the Hunchback repressor, which is localized in the anterior half of the embryo. Hunchback disappears at gastrulation (3–4 h after fertilization), but repression of *Ubx* in the anterior half of the embryo is maintained by the PcG silencing mechanism. PcG proteins are present from the earliest stages since they are maternally supplied in the oocyte but they do not repress the initial activation of the *Ubx* gene. We may distinguish therefore three stages in the establishment of the *Ubx* pattern of expression (Chan et al. 1993; Poux et al. 1996).

Stage 1 or the naïve stage, 0–3 h after fertilization. Expression begins wherever activators are present and the Hunchback repressor is not present. PcG proteins do not interfere.

Stage 2 or the fixation stage, 3–5 h after fertilization. Around gastrulation silencing of *Ubx* occurs if the gene is not active.

Stage 3 or the maintenance stage, during later embryonic and postembryonic development. The silenced state of *Ubx* is maintained in the cellular progeny. In cells where *Ubx* is not silenced (and their progeny), it remains competent to be transcribed during later development whenever activators are present.

6.3 The Polycomb Group Proteins

The PcG genes do not constitute a family but a diverse group originally defined by the fact that loss of function of each member has similar consequences: derepression of the homeotic genes. Molecular analysis has shown that most PcG gene products act together forming multiprotein complexes. Biochemical studies have identified two basic kinds of complexes with distinct functions that must cooperate to produce gene silencing. The two types are frequently referred to as PRC1 and PRC2 but both appear to have many subspecies with somewhat different composition. PRC1 includes a core quartet of proteins: Polycomb (PC), Polyhomeotic (PH), Posterior sex combs (PSC) and dRing (Saurin et al.

2001). Each of these proteins contains structural domains suggestive of possible functions. PC contains a chromodomain, a feature also found in the heterochromatin protein HP1 and now known to bind to methylated lysines in the N-terminal tail of histone H3. PSC and dRing both contain RING domains, found in many proteins that act as E3 ubiquitin ligases. The PRC2 complex contains E(Z), a SET domain protein responsible for histone H3 methyltransferase activity, as well as ESC and p55, both WD40 domain proteins (Czermin et al. 2002; Müller et al. 2002; Cao et al. 2002; Kuzmichev et al. 2002). The p55 protein is the ortholog of mammalian RbAp46/48 and is also a component of the chromatin assembly complex. Both kinds of complexes have been found associated with the histone deacetylase Rpd3. Of the two, PRC2 appears to be phylogenetically the most ancient since homologs of E(Z) and ESC exist and are known to act together in plants as well as in animals.

6.4 Polycomb Response Elements

The regulatory sequences that mediate PcG responses are the Polycomb response elements (PREs). The PREs that have been identified, for the most part in homeotic gene complexes, are regions of several hundred base pairs that can be located upstream, within, or downstream of a transcription unit. In the *Ubx* gene, the major PRE lies 24 kb upstream of the promoter, while a second, weaker PRE is within an intron 30 kb downstream from the transcription start. Transgene constructs containing all or parts of a PRE region reveal PcG-dependent silencing effects that can act upon multiple genes placed in the vicinity of the PRE. In addition, the PREs studied contain sequences that respond to Trithorax, another SET domain protein whose histone H3 methyltransferase activity is specific for lysine 4 (Czermin et al. 2002; Smith et al. 2004).

Both kinds of PcG complexes, PRC1 and PRC2, are recruited to the PRE, which must therefore contain specifically recognized sequences. Yet, the PcG proteins themselves do not have specific DNA binding activities and PREs from different genes do not display obvious sequence similarities. Comparison of many PREs and of a given PRE in different *Drosophila* species shows the presence of multiple copies of several short sequence motifs, some of which are recognizable consensus sequences for DNA-binding proteins such as GAGA factor, PHO (the fly homolog of mammalian YY1), Zeste or DSP1 (Horard et al. 2000; Dellino et al. 2002; Ringrose et al. 2003; Déjardin et al. 2005). Each of these proteins binds also to many other genomic sites, including many promoters that have no PRE function. These considerations, together with the functional dissection of PREs, suggest that the PRE-binding proteins act cooperatively and in partly redundant fashion to recruit the PRC1 and PRC2 complexes. Although the details of the recruiting process remain to be elucidated, experiments using protein fusions with GAL4 or LexA DNA-binding domains show that, in vivo, the stable binding of one PcG protein suffices to recruit a functional silencing complex (Müller 1995; Poux et al. 2001a,b).

A major difficulty in the study of mammalian PcG mechanisms is that no functional mammalian PRE has yet been identified. Despite the conservation of PcG proteins from flies to vertebrates, none of the proteins thought to be recruiters, with the exception of PHO/YY1, are conserved in mammalian systems.

6.5 PRE Properties

The addition of core PRE sequences to a reporter gene is generally sufficient to induce PcG-dependent silencing of that gene and often of other genes in the physical proximity. Two basic aspects of the silenced state conferred by the PRE are its inheritability and its all-or-nothing nature. A gene silenced by the action of a neighboring PRE will tend to remain silenced in subsequent cell cycles, thereby transmitting the silenced state to the cellular progeny in an epigenetic fashion. The all-or-none quality is illustrated by the fact that decreasing the levels of PcG proteins increases the probability of losing the silenced state, giving rise to clones of cells in which the target gene is derepressed. Thus, silencing of the *white* gene, necessary for eye pigmentation, gives rise to characteristically variegated eyes, highly reminiscent of position-effect variegation seen when the *white* gene is relocated to the vicinity of heterochromatin.

A characteristic feature of the effects mediated by the PRE is that they are often strongly enhanced by the pairing of homologous chromosomes, each carrying the PRE-containing transgene. This characteristic, often called pairing-dependent silencing, might be attributed to the increased local concentration of PcG proteins and consequent greater stability of the repressed state.

The PcG proteins, produced from maternal transcripts deposited during oogenesis, are present in the embryo at very early stages but they do not prevent the initial activation of homeotic genes. At later stages, they are nearly ubiquitously expressed, but they act differentially on their target genes to allow continued expression where expression was initially activated while maintaining transcriptional silence in those cells in which the target genes were repressed. Thus the PcG mechanism must sense the state of activity of its targets in the blastoderm stage embryo and determine accordingly whether to silence or to allow continued expression.

The PREs so far analyzed have been found to respond to Trithorax control, as well as to PcG control. That is, *trx* mutations reduce the level of expression of the gene regulated by the PRE and can cause the return of silencing even after early transcriptional activity. The functional dissection of these PREs indicates that they contain a Trithorax response element that can be mutated independently without affecting the silencing activity (Tillib et al. 1999). In this respect, the PRE contains two functions, one repressive and dependent on the PcG proteins and another that stimulates expression and depends on Trithorax. Both functions behave as if they had an associated memory: once repression is established, it is maintained by a "negative" memory that prevents later activation, while if the target gene was initially active, repression is not established and the nonrepressed state is maintained by a "positive" memory that prevents later silencing by the ubiquitous PcG proteins (Poux et al. 2002; Klymenko and Müller 2004). The combination of these two antagonistic effects, both mediated by the homeotic PREs, may be responsible for the all-or-none nature of PcG silencing.

6.6 Chromatin Modifications

The similarity between the effects of PcG silencing and those of heterochromatic position-effect variegation led to the commonly held belief that they have similar mechanisms. The transcriptional silencing was thought to derive from the spread of PcG complexes by continuous self-assembly to cover a chromatin domain encompassing one or more genes. The complexes were supposed to condense the chromatin into a tightly packaged form inaccessible to transcription factors or RNA polymerase. This view was encouraged by the model of the yeast Sir complexes. The Sir proteins are first recruited at the telomeric repeats and then spread over many kilobases by an assembly process mediated by interactions between Sir proteins and deacetylated nucleosomes (Hecht et al. 1996; Rusche et al. 2003). In *Drosophila*, however, chromatin immunoprecipitation experiments have not shown extensive spreading of PcG complexes. In the Bithorax complex, PcG proteins appear to be primarily localized at or around the known PREs, although a low level distinctly higher than background is found over a broader region (see for example Ringrose et al. 2004).

The parallel between heterochromatin complexes and PcG complexes is nevertheless a real one in some respects. Both have important histone modifying activities. Heterochromatin complexes methylate histone H3 at lysine 9 through the action of Su(var)3-9 (Rea et al. 2000; Schotta et al. 2004). In PcG complexes, a similar role is played by E(Z), which methylates H3 lysine 9 and 27 (Czermin et al. 2002; Müller et al. 2002; Cao et al. 2002; Kuzmichev et al. 2002). In heterochromatin, the di- or trimethylated lysine 9 is detected by the chromodomain of HP1 (Bannister et al. 2001; Lachner et al. 2001). The chromodomain of the Polycomb protein has been shown to interact specifically with H3 trimethylated at lysine 27 (Czermin et al. 2002; Fischle et al. 2003). The histone deacetylase Rpd3, associated with both kinds of PRE complexes, presumably deacetylates histone H3 to make the lysines available for methylation. The specific methylation of the nucleosomes has been considered a good candidate for the epigenetic mark that facilitates the reconstitution of PcG complexes every cell cycle and therefore serves as the memory of the silenced state.

Recently, another histone modification has been found to be produced by PcG complexes. Ubiquitylation of histone H2A is mediated in vitro by dRing, a core component of the Polycomb complex, and is found in vivo at PcG silenced loci (H. Wang et al. 2004). The role of this modification is still unclear. We might speculate that, as ubiquitylation of histone H2B is a prerequisite for the H3 lysine 4 and lysine 79 methylation that is associated with transcriptional elongation (Henry et al. 2003), ubiquitylation of H2A might serve a symmetric role to permit the methylation of lysine 27. However, the mechanistic aspects of these relationships are not yet understood.

The antagonistic epigenetic mark, signifying the derepressed state is thought to be conferred by Trithorax, which is also a SET domain protein with histone methyltransferase activity. Trithorax methylates H3 lysine 4, a chromatin mark that is known to be associated with transcribed regions. It is not clear how the two epigenetic marks antagonize one another.

One possible explanation for the bimodal all-or-none function of the PRE might depend on transcriptional activity through the PRE itself. Schmitt et al. (2005) have shown in fact that silencing can be abolished by driving transcription across the PRE. The regulatory regions of the homeotic genes contain numerous, apparently noncoding transcription units, generally traversing known PRE regions. In this view, incipient derepression of the PRE would result in transcriptional activity across the PRE itself. Transcriptional elongation is known to comport a variety of chromatin modification activities, including histone acetylation and histone H3 lysine 4 and 79 methylation, which could require Trithorax activity. These modifications may be incompatible with the formation or function of PcG complexes and therefore would result in loss of silencing and a complete switch to the derepressed state. While this would certainly produce an all-or-none response, switch activity has also been found in other cases where no transcription across the PRE can be detected (Déjardin and Cavalli, 2004).

6.7 Histone Methylation and Recruitment

The finding that the chromodomain of PC has a specific affinity for H3 lysine 27 has encouraged the hypothesis that E(z) complexes are recruited first to the PRE, methylate the nucleosomes and ipso facto recruit the Polycomb complex by affinity with the methylated site (Cao et al. 2002; L. Wang et al. 2004). This model, analogous to a similar model proposed for HP1 binding to heterochromatin, is probably not correct. The affinities of PC or HP1 for their corresponding methylated peptides, while significant, are at least three orders of magnitude lower than those typically found for DNA-binding proteins and their specific binding sequences (Fischle et al. 2003). Targeted methylation of chromatin sites does not appear to automatically recruit the corresponding chromodomain protein (Stewart et al. 2005). When H3 methylation is lost, for example, in the case of a thermosensitive mutant E(Z), the loss of PcG complex binding to PRE sites lags far behind the loss of detectable methylation (Czermin et al. 2002). Finally, our mapping of Polycomb binding relative to lysine 27 trimethylation, using chromatin immunoprecipitation quantitated by real-time PCR, shows that methylation occurs over a broad domain (over 100 kb of the *Ubx* gene), while the PcG proteins are primarily (though not exclusively) localized over the PRE (Y. Schwartz, T. Kahn and V. Pirrotta). Nevertheless, histone methylation is essential for stable binding of PcG complexes to their target genes. A possible explanation is that the large methylated chromatin domain helps to retain a local concentration of Polycomb to ensure occupancy of the PRE.

6.8 The Silencing Mechanism

The classical interpretation of heterochromatic silencing was that heterochromatin proteins condensed and packaged chromatin into higher-order structures that rendered the DNA inaccessible to activators and transcriptional machinery. This general sense was applied also to PcG silencing and is still widely held today. Various attempts to show that PcG silencing inhibits access to nucleases or DNA methyltransferases have given conflicting results and have not generally supported this model. While some decrease in nuclease accessibility has been found in some studies (Boivin and Dura 1998), it does not seem to be necessary or sufficient to account for silencing (see Dellino et al. 2004). In addition, the PRE itself is highly sensitive to nucleases and restriction enzymes. A study of a heat shock promoter construct silenced by a PRE (Dellino et al. 2004) showed no major changes in nucleosome configuration and in the presence of TFIID and RNA pol II, which are normally constitutively recruited to the heat shock promoter. There was also no interference in the binding of heat shock factor upon heat treatment. When the promoter was silenced by the vicinity of a PRE, however, the RNA polymerase was unable to open the strands and initiate transcription. This strongly suggests that PcG silencing affects the function directly, not the assembly of the general transcription factors at the heat shock promoter. These results do not exclude the possibility that PcG complexes might also have effects antagonistic to other processes required for transcription at other kinds of promoter. Some promoters, for example, require chromatin remodeling before they become accessible to activators or to general transcription factors. Interference with remodeling could block transcription at an earlier stage. The experiments with the heat shock promoter make it very unlikely that silencing is the consequence of chromatin packaging or condensation.

6.9 Alternative Modes of Action

The remainder of this discussion is devoted to the possibility that the standard model for PcG action, based on the homeotic gene paradigm, is not the only mode of action of PcG complexes. In particular, two characteristics of PcG action in the context of homeotic gene regulation – the all-or-none aspect and the long-term stability of the silenced state – are not necessarily a universal feature of mechanisms based on PcG proteins.

We understand the role of very few PREs outside of those found in homeotic genes. There are some cases, however, where it is unlikely that PcG mechanisms function in an all-or-none silencing mode. Two such cases are the *polyhomeotic* gene (*ph*) and the *Posterior sex combs* (*Psc*) gene, homologs of mammalian Rae28 and Bmi1, respectively. Both genes are themselves members of the PcG. Their regulatory regions are also binding sites for PcG complexes, suggesting that they are autoregulated by a PRE-like mechanism but a complete auto-silencing of these genes is unlikely since their products are needed for the silencing itself. In vivo evidence, however, indicates that at certain stages they do become down-regulated in a manner dependent on PcG genes (Fauvarque et al. 1995; Rastelli et al. 1993).

The *ph* PRE has been analyzed is some detail (Fauvarque and Dura 1993; Bloyer et al. 2003) and presents unexpected features. When placed in front of a reporter gene, it can cause pairing-dependent repression or variegation of the *white* gene in reporter constructs, but, unlike other

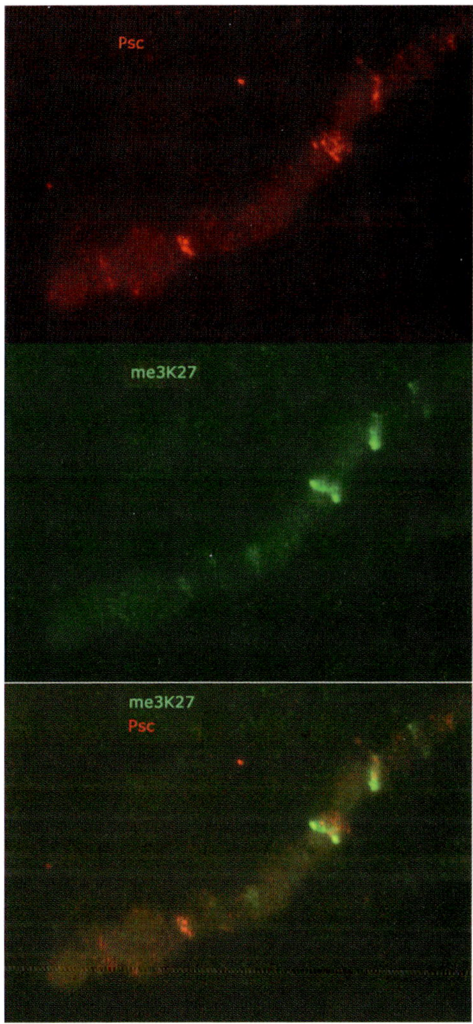

Fig. 1. Some chromosomal PcG target sites are undermethylated. The tip of the X salivary gland polytene chromosome was doubly immunostained with antibody against histone H3 me3K27 (*green*) and the PcG protein PSC (*red*). One site that stains strongly with anti-PSC but very weakly with anti-H3 me3K27 is the site of the PcG gene polyhomeotic

PREs, it effects this repression only when placed close to the promoter. In addition, in some contexts decreasing the dosage of certain PcG genes such as *Pc* or *Psc* have an effect contrary to that expected: they decrease expression of the reporter gene, as if these products were necessary to stimulate expression. Such paradoxical responses are known in the case of heterochromatic silencing. The expression of certain *Drosophila* genes that normally reside in heterochromatin is not only not silenced by the heterochromatic environment but is in fact dependent on HP1 and other heterochromatin proteins. Another feature of both the *ph* and the *Psc* locus visible on polytene chromosomes is that while they stain strongly with antibodies against components of the Polycomb complex, they stain only weakly with anti-H3 me3K27, relative to other PcG binding sites (see Fig. 1 and Czermin et al. 2002). This observation argues against the idea that K27 methylation directly recruits PcG complexes and suggests the possibility that it may instead be related to the degree of silencing effected. The *ph* PRE lies immediately upstream of the promoter and no transcripts are known to traverse the *ph* regulatory regions. In the transcription-mediated switch model, we might suppose that incipient derepression of the *ph* PRE would not activate transcription across the PRE and therefore, rather than automatically switching off silencing, the promoter might be regulated in a continuously graded manner. At the same time, the transcriptional activity would antagonize the PcG-induced H3 lysine 27 methylation and produce instead methylation of lysine 4 and 79 along the transcribed region.

6.10 Reprogramming

According to the homeotic paradigm, once the silenced state is established, it is stably transmitted to progeny cells for the rest of development. Conversely, once the nonrepressed state is established (by transcriptional activity at early stages) the interaction between PcG and trxG mechanisms prevents subsequent PcG silencing. Here I want to explore possible means of reprogramming PcG silencing by reactivation or by late-onset silencing.

Derepression of a gene silenced at an earlier stage could occur by interfering with the PRE silencing function. In the homeotic genes, early transcriptional activity establishes a state antagonistic to the recruitment of PcG complexes. This probably occurs through the histone acetylation and H3 K4 and K79 methylation that are associated with transcriptional activity and with the Trithorax complex, although the exact role of Trx at the PRE remains unclear. A repressed state can be lost by massive expression of an activator such as GAL4 targeted to a nearby promoter region, even when transcription does not traverse the PRE (Cavalli and Paro 1998, 1999; Déjardin and Cavalli 2004). It is possible that certain activators are more effective at overcoming PcG silencing and are employed to reactivate repressed genes. An example of this reactivation may occur at the *engrailed* gene in certain domains of the wing disc where it had previously been silenced (Maurange and Paro 2002).

Another way to block silencing activity is by the action of a chromatin insulator (Sigrist and Pirrotta 1997). The insertion of the insulator between a PRE and a target promoter completely protects the promoter from silencing. Furthermore, the insulator also blocks the spread of H3 methylation into the reporter gene (Y. Schwartz, T. Khan, V. Pirrotta, unpublished data). Therefore, the targeting of insulator proteins to a potential insulator could be used to block silencing and derepress a target gene.

6.11 Export of PcG Proteins

Recent reports indicate that, in mammalian cells, at least some PcG proteins can be exported out of the nucleus after activation of certain signal transduction cascades. We do not know yet the physiological significance of this effect but activation of certain integrins (Rietzler et al. 1998; Witte et al. 2004) or of T cell receptor (Hobert et al. 1996; Ogawa et al. 2003) cause both Eed (the mammalian homolog of *Drosophila* ESC) and Ezh2 [E(z)] to shuttle from the nucleus to the cytoplasm and associate with components of a large multiprotein complex that mediates signal transduction. It is not known if Ezh2 plays a methyltransferase role in this complex but its migration out of the nucleus raises very interesting possibilities. Global depletion of the Ezh2/Eed complex from the nucleus would be expected to result in eventual derepression of all PcG targets, hence globally reset the state of PcG silenced genes.

More interestingly, a similar but more specific mechanism could be used to deplete the complex from specific chromatin sites. Phosphorylation of at least some PcG proteins accompanies their dissociation from chromatin at mitosis (Voncken et al. 1999). Both E(Z) and ESC are known to become phosphorylated, although the significance has not ben studied, it is likely that this is related to their cytoplasmic presence. Global phosphorylation would cause global derepression, but activation of a specific kinase that binds only at specific genomic sites could produce specific derepression and effectively reprogram a specific subset of PcG target loci. The identification of such specific kinases or the design of a kinase that can be specifically targeted to a desired PcG site would open the way to the reprogramming of a desired chromatin domain.

References

Bannister AJ, Zegerman P, Partridge JF, Miska EA, Thomas JO, Allshire RC, Kouzarides T (2001) Selective recognition of methylated lysine 9 on histone H3 by the HP1 chromo domain. Nature 410:120–124

Bloyer S, Cavalli G, Brock HW, Dura J-M (2003) Identification and characterization of polyhomeotic PREs and TREs. Dev Biol 261:426–442

Boivin A, Dura J-M (1998) In vivo chromatin accessibility correlates with gene silencing in *Drosophila*. Genetics 150:1539–1549

Cao R, Wang L, Wang H, Xin L, Erdjument-Bromage H, Tempst P, Jones RS, Zhang Y (2002) Role of histone H3 lysine 27 methylation in Polycomb-Group silencing. Science 298:1039–1043

Cavalli G, Paro R (1998) The *Drosophila Fab-7* chromosomal element conveys epigenetic inheritance during mitosis and meiosis. Cell 93:505–518

Cavalli G, Paro R (1999) Epigenetic inheritance of active chromatin after removal of the main transactivator. Science 286:955–958

Chan C-S, Rastelli L, Pirrotta V (1994) A Polycomb response element in the *Ubx* gene that determines an epigenetically inherited state of repression. EMBO J 13:2553–2564

Czermin B, Melfi R, McCabe D, Seitz V, Imhof A, Pirrotta V (2002) *Drosophila* enhancer of Zeste/ESC complexes have a histone H3 methyltransferase activity that marks chromosomal Polycomb sites. Cell 111:185–196

Déjardin J, Cavalli G (2004) Chromatin inheritance upon Zeste-mediated Brahma recruitment at a minimal cellular memory module. EMBO J 23:857–868

Déjardin J, Rappailles A, Cuvier O, Grimaud C, Decoville M, Locker D, Cavalli G (2005) Recruitment of *Drosophila* Polycomb group proteins to chromatin by DSP1. Nature 434:533–538

Dellino GI, Tatout C, Pirrotta V (2002) Conservation of sequences and chromatin structure of the *bxd* Polycomb response element in *Drosophila* species. Int J Dev Biol 46:133–141

Dellino GI, Schwartz YB, Farkas G, McCabe D, Elgin SCR, Pirrotta V (2004) Polycomb silencing blocks transcription initiation. Mol Cell 13:887–893

Fauvarque M-O, Dura J-M (1993) *polyhomeotic* regulatory sequences induce developmental regulator-dependent variegation and targeted P-element insertions in *Drosophila*. Genes Dev 7:1508–1520

Fauvarque M-O, Zuber V, Dura J-M (1995) Regulation of *polyhomeotic* transcription may involve local changes in chromatin activity in *Drosophila*. Mech Dev 52:343–355

Fischle W, Wang Y, Jacobs SA, Kim Y, Allis CD, Khorasanizadeh S (2003) Molecular basis for the discrimination of repressive methyl-lysine marks in histone H3 by Polycomb and HP1 chromodomains. Genes Dev 17:1870–1881

Hecht A, Strahl-Bolsinger S, Grunstein M (1996) Spreading of transcriptional repressor SIR3 from telomeric heterochromatin. Nature 383:92–96

Henry KW, Wyce A, Lo W-S, Duggan LJ, Emre NCT, Kao C-F, Pillus L, Shilatifard A, Osley MA, Berger SL (2003) Transcriptional activation via sequential histone H2B ubiquitylation and deubiquitylation, mediated by SAGA-associated Ubp8. Genes Dev 17:2648–2663

Hobert O, Jallal B, Ullrich A (1996) Interaction of Vav with ENX-1, a putative transcriptional regulator of homeobox gene expression. Mol Cell Biol 16:3066–3073

Horard B, Tatout C, Poux S, Pirrotta V (2000) Structure of a polycomb response element and in vitro binding of polycomb group complexes containing GAGA factor. Mol Cell Biol 20:3187–3197

Klymenko T, Müller J (2004) The histone methyltransferases Trithorax and Ash1 prevent transcriptional silencing by Polycomb group proteins. EMBO Reports 5:373–377

Kuzmichev A, Nishioka K, Erdjument-Bromage H, Tempst P, Reinberg D (2002) Histone methyltransferase activity associated with a human multiprotein complex containing the Enhancer of Zeste protein. Genes Dev 22:2893–2905

Lachner M, O'Carroll D, Rea S, Mechtler K, Jenuwein T (2001) Methylation of histone H3 lysine 9 creates a binding site for HP1 proteins. Nature 410:116–120

Maurange C, Paro R (2002) A cellular memory module conveys epigenetic inheritance of hedgehog expression during *Drosophila* wing imaginal disc development. Genes Dev 16:2672–2683

Müller J (1995) Transcriptional silencing by the Polycomb protein in *Drosophila* embryos. EMBO J 14:1209–1122

Müller J, Hart CM, Francis NJ, Vargas ML, Sengupta A, Wild B, Miller EL, O'Connor MB, Kingston RE, Simon JA (2002) Histone methyltransferase activity of a *Drosophila* Polycomb Group repressor complex. Cell 111:197–208

Ogawa M, Hiraoka Y, Aiso S (2003) The Polycomb-group protein ENX-2 interacts with ZAP-70. Immunol Lett 86:57–61

Poux S, Kostic C, Pirrotta V (1996) Hunchback-independent silencing of late *Ubx* enhancers by a Polycomb Group Response Element. EMBO J 15:4713–4722

Poux S, McCabe D, Pirrotta V (2001a) Recruitment of components of Polycomb Group chromatin complexes in *Drosophila*. Development 128:75–85

Poux S, Melfi R, Pirrotta V (2001b) Establishment of Polycomb silencing requires a transient interaction between PC and ESC. Genes Dev 15:2509–2514

Poux S, Horard B, Sigrist CJA, Pirrotta V (2002) The *Drosophila* Trithorax protein is a coactivator required to prevent re-establishment of Polycomb silencing. Development 129:2843–2893

Rastelli L, Chan CS, Pirrotta V (1993) Related chromosome binding sites for zeste, suppressors of zeste and Polycomb group proteins in *Drosophila* and their dependence on Enhancer of zeste function. EMBO J 12:1513–1522

Rea S, Eisenhaber F, O'Carroll D, Strahl BD, Sun Z-W, Schmid M, Opravil S, Mechtler K, Ponting CP, Allis CD, Jenuwein T (2000) Regulation of chromatin structure by site-specific histone H3 methyltransferases. Nature 406:593–599

Rietzler M, Bittner M, Kolanus W, Schuster A, Holzmann B (1998) The human WD repeat protein WAIT-1 specifically interacts with the cytoplasmic tails of beta7-integrins. J Biol Chem 273:27459–27466

Ringrose L, Rehmsmeier M, Dura JM, Paro R (2003) Genome-wide prediction of Polycomb/Trithorax Response Elements in *Drosophila* melanogaster. Dev Cell 5:759–771

Ringrose L, Ehret H, Paro R (2004) Distinct contributions of histone H3 Lysine 9 and 27 methylation to locus-specific stability of Polycomb complexes. Mol Cell 16:641–653

Rusche LN, Kirchmaier AL, Rine J (2003) The establishment, inheritance, and function of silenced chromatin in *Saccharomyces cerevisiae*. Ann Rev Biochem 72:481–516

Saurin AJ, Shao Z, Erdjument-Bromage H, Tempst P, Kingston RE (2001) A *Drosophila* Polycomb group complex includes Zeste nd dTAFII proteins. Nature 412:655–660

Schmitt S, Prestel M, Paro R (2005) Intergenic transcription through a Polycomb group response element counteracts silencing. Genes Dev 19:697–708

Schotta G, Lachner M, Sarma K, Ebert A, Sengupta R, Reuter G, Reinberg D, Jenuwein T (2004) A silencing pathway to induce H3-K9 and H4-K20 trimethylation at constitutive heterochromatin. Genes Dev 18:1251–1262

Sigrist CJA, Pirrotta V (1997) Chromatin insulator elements block the silencing of a target gene by the *Drosophila* Polycomb Response Element (PRE) but allow trans interactions between PREs on different chromosomes. Genetics 147:209–221

Smith ST, Petruk S, Sedkov Y, Cho E, Tillib S, Canaani E, Mazo A (2004) Modulation of heat shock gene expression by the TAC1 chromatin-modifying complex. Nat Cell Biol 6:162–167

Stewart MD, Li J, Wong J (2005) Relationship between Histone H3 Lysine 9 Methylation, Transcription Repression, and Heterochromatin Protein 1 Recruitment. Mol Cell Biol 25:2525–2538

Tillib S, Petruk S, Sedkov Y, Kuzin A, Fujioka M, Goto T, Mazo A (1999) Trithorax- and Polycomb-Group Response Elements within an Ultrabithorax transcription maintenance unit consist of closely situated but separable sequences. Mol Cell Biol 19:5189–5202

Voncken JW, Schweizer D, Aagaard L, Sattler L, Jantsch MF, van Lohuizen M (1999) Chromatin association of the Polycomb group protein BMI1 is cell cycle-regulated and correlates with it phosphorylation status. J Cell Sci 112:4627–4639

Wang H, Wang L, Erdjument-Bromage H, Vidal M, Tempst P, Jones RS, Zhang Y (2004) Role of histone H2A ubiquitination in Polycomb silencing. Nature 431:873–878

Wang L, Brown JL, Cao R, Zhang Y, Kassis JA, Jones RS (2004) Hierarchical recruitment of Polycomb Group silencing complexes. Mol Cell 14:637–646

Witte V, Laffert B, Rosorius O, Lischka P, Blume K, Galler G, Stilper A, Willbold D, D'Aloje P, Sixt M et al. (2004) HIV-1 Nef mimics an integrin receptor signal that recruits the Polycomb Group protein Eed to the plasma membrane. Mol Cell 13:179–190

7 CpG Island Methylation and Histone Modifications: Biology and Clinical Significance

M. Esteller

Abstract. The discovery that drastic changes in DNA methylation and histone modifications are common in human tumors has inspired various laboratories and pharmaceutical companies to develop and study epigenetic drugs. One of the most promising groups of agents is the inhibitors of histone deacetylases (HDACs), which have different biochemical and biologic properties but have a single common activity: induction of acetylation in histones, the key proteins in nucleosome and chromatin structure. HDAC inhibitors may act through the transcriptional reactivation of dormant tumor-suppressor genes. However, their pleiotropic nature leaves open the possibility that their well-known differentiation, cell-cycle arrest, and apoptotic properties are also involved in other functions associated with HDAC inhibition. Many phase I clinical trials indicate that HDAC inhibitors appear to be well-tolerated drugs. Thus, the field is ready for rigorous biologic and clinical scrutiny to validate the therapeutic potential of these drugs. HDAC inhibitors, probably in association with classical chemotherapy drugs or in combination with DNA-demethylating agents, could be promising drugs for cancer patients.

7.1 Introduction

During the past few years, chromatin has revealed its essential role in controlling gene activity. Although the first studies in chromatin structure suggested that its major role could be the packaging of the genome in the cell nucleus, most recent evidence has led to a novel picture in which chromatin is also a very dynamic entity where histone post-translational modifications play a key role in modulating gene expression.

Chromatin has also moved to a key position in cancer research as epigenetic alterations and different types of chromatin changes have been revealed to occur in cancer cells. One of the first identified epigenetic alterations in cancer cells is the imbalance of DNA methylation patterns (Jones and Baylin 2002). DNA methylation changes occur both in CpG islands, where CpG hypermethylation results in gene silencing, and in dispersed CpGs that suffer hypomethylation (Feinberg and Tycko 2004). DNA methylation changes are directly associated with histone modification changes at both CpG islands (Ballestar et al. 2003) and repetitive sequences (Fraga et al. 2005).

Although DNA methylation and histone modification changes are generally associated, there are also methylation-independent histone modification alterations in cancer cells. An archetypical example is provided by the loss of expression of p21 that is associated with histone deacetylation (Archer et al. 1998; Blagosklonny et al. 2002; Gui et al. 2004). Another source of histone modification alterations is provided by the existence of translocations resulting in abnormal proteins that contain a histone-modifying activity fused to another protein that mistargets the activity to an aberrant DNA sequence.

An important issue concerning epigenetic alterations in the context of a cancer cell, when compared to genetic lesions, is the possibility of reversion when using appropriate inhibitors of the DNA methylation and histone modification machineries.

7.2 Histone Modification Directs Chromatin Structure and Function

The histone octamer, which constitutes the protein core of the nucleosome, the repetitive subunit of the chromatin, provides a smart solution as it facilitates the formation of a fibber that is capable to enter into several additional levels of compaction. Core histones possess long N-terminal ends, spanning about one-third of their total length, which protrude outside the nucleosome. Histone N-terminal tails are a key element in chromatin dynamics. The core histone N-terminal tails constitute an excellent focal point for interactions with the myriad nuclear factors that will read the signals encoded by post-translational modifications occurring in their highly conserved K, R and S residues and participate in nucleosomal dynamics.

Many different histone-modifying enzymes alter the post-translational modification status of histone residues and contribute to regulation of the expression of DNA sequences underneath. In particular, histone acetyltransferases (HATs) and deacetylases (HDACs) (Fig. 1), involved in the balance of histone acetylation, and different histone methyltransferases (HMTs), which catalyze methylation at K or R residues, are the most studied.

The current model of how histone modifications are used by the cell, known as the histone code hypothesis, proposes that the combinatorial and/or sequential post-translational modification of the histone residues can be read by nuclear factors to promote different processes (Strahl and Allis 2000; Turner 2000). The specificity of the post-translational modifications of histones can be appreciated, for instance, in the methylation of H3 at K4 and R17, closely linked to transcriptional competence or the methylation of H3 at K9 or H4 at K20, associated with transcriptional repression (Turner 2005; Peterson and Laniel 2004). Similarly to the studies on gene function in the 1980s that evidenced the need for a detailed map of the genome, the future of epigenetics requires a detailed map of epigenetic modifications at a genomic scale. Although H3-K9 and H4-K20 trimethylation appears stable at several heterochromatic

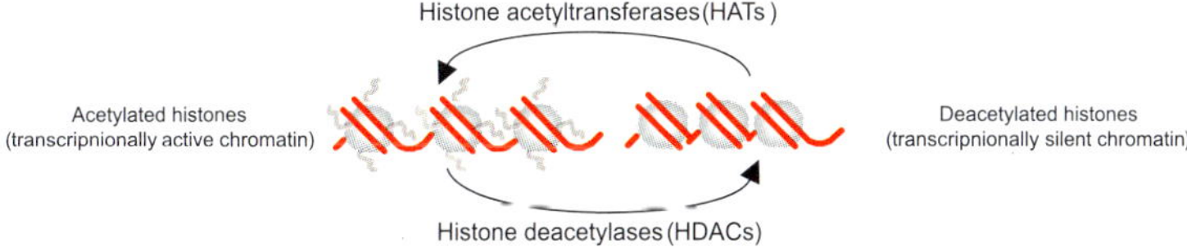

Fig. 1. Acetylation-mediated modification of chromatin by particular histone acetyltransferases (*HATs*) and deacetylases (*HDACs*)

locations, significant variation is found depending on the type of cells (undifferentiated and differentiated ES cells, embryonic trophoblasts and fibroblasts) (Martens et al. 2005).

The dissection of the histone code also needs the dissection of the machinery involved in its establishment. Histone acetylation and phosphorylation are obvious transient signals, as the enzymatic activities that introduce and revert this modification are well known. For years, histone methylation has appeared as a more stable signal. A third issue, besides the mapping and identification of histone modification enzymes, in deciphering the histone code that needs to be addressed is the definition of the histone code.

7.3 Alteration of the Histone Modification Patterns in Cancer Cells

The transformation from normal to cancer cells involves multiple etiologic pathways and, at the very least, is a genetic, epigenetic and cytogenetic process. Until recently, the genetic route to tumorigenesis was the most studied; however, epigenetics is now coming of age, and most novel research is taking place in this emerging area. DNA methylation, especially methylation-associated silencing of tumor-suppressor genes, is the most widely analyzed type of epigenetic alteration of human tumors (Esteller 2002). However, another extremely important epigenetic route to carcinogenesis involves the aberrant pattern of histone modifications. To understand this, we need to look at the ultrastructural level, concentrating our attention on the nucleosomes, which are key regulators of gene expression. Nucleosomes contain 146 DNA base pairs wrapped around eight histone subunits: two copies each of H2A, H2B, H3 and H4. These proteins have an amino-terminal tail rich in the amino acid lysine, which can undergo different post-translational modifications: acetylation, methylation, phosphorylation and ubiquitination, among others. These modifications are read by different proteins and complexes involved in chromatin remodeling and transcriptional activation or repression (Jenuwein and Allis 2001; Iizuka and Smith 2003). There are many combinations of histone modifications that may constitute a proposed histone code (Jenuwein and Allis 2001; Iizuka

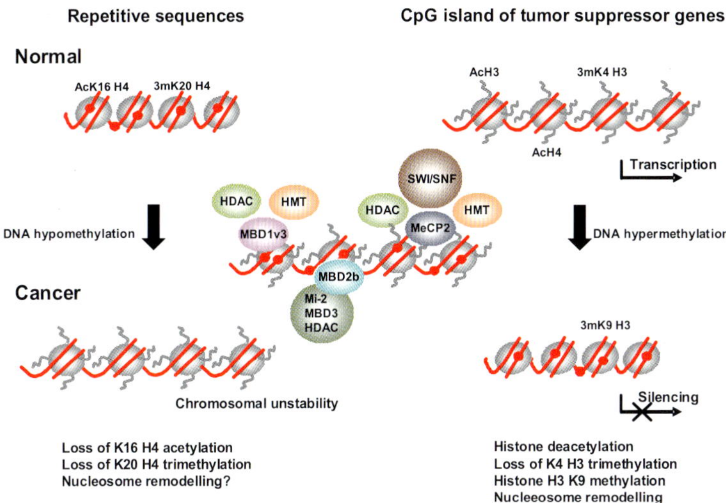

Fig. 2. Histone modification changes in cancer cells. The patterns of histone modifications in repetitive sequences (satellite/nonsatellite) (*left*) and the CpG island of tumor suppressor genes (*right*) are shown. Repetitive sequences suffer global DNA hypomethylation, while CpG islands undergo hypermethylation. In this scenario, DNMTs and MBDs couple DNA methylation changes to histone modification alterations. Arrays of nucleosomes in the two groups of regions are shown. Histone octamers are represented by *grey circles*. DNA is represented as a *red line* in which only methylated CpG dinucleotides are shown as *red circles*. Acetylated histone tails are *protruding lines* from octamers, whereas deacetylated histone tails are not shown. In the normal cell (*top*), CpG islands at the promoter of tumor suppressor genes are unmethylated and these genes are transcriptionally active. In cancer cells, many tumor-suppressor genes undergo aberrant hypermethylation at their CpG islands and many different elements can be recruited: MBD proteins involved in transcriptional repression, recruitment of HDAC, HMTs, and remodeling complexes (SWI/SNF, Mi-2), binding to both methyl-CpG-rich and -poor sequences as well as modulation of the binding by post-translational modification. Loss of methylation in repetitive sequences and subsequent histone modification alterations are associated with chromosomal instability

and Smith 2003). One of the best-studied histone modifications is acetylation of some of the lysine residues of histones H3 and H4 (Fig. 1). This is reversible and carefully controlled by two large groups of enzymes: HATs and HDACs. The presence of certain acetylated lysines within the histone tails is associated with less condensed chromatin and a transcriptionally active gene status, whereas the deacetylated residues are associated with heterochromatin and transcriptional gene silencing (Jenuwein and Allis 2001; Iizuka and Smith 2003; Johnstone 2002).

Cancer cells undergo multiple alterations of the histone code (Fig. 2). Promoter hypermethylation of tumor suppressor genes is generally linked to changes in the histone modification pattern. Some of these changes in histone modifications have been characterized during the past few years. For instance, most genes that are hypermethylated in cancer are hypoacetylated at histones H3 and H4. Moreover, in some cases K9 methylation has been demonstrated to occur in hypermethylated DNA sequences. Usually DNA aberrant hypermethylation is accompanied by loss of K4 trimethylation.

Apart from the alterations of the histone code at the promoter region of tumor suppressor genes in cancer, today there is more and more evidence that cancer cells also present an altered pattern of histone modifications within constitutive heterochromatin. For instance, it has been recently found that cancer cells had a loss of monoacetylated and trimethylated forms of histone H4 and that these changes appeared early and accumulated during the tumorigenic process (Fraga et al. 2005). These alterations of the normal profile of histone modification occurred predominantly at the acetylated Lys16 and trimethylated Lys20 residues of histone H4 and are associated with the hypomethylation of DNA repetitive sequences, a well-known characteristic of cancer cells. These data suggest that the global loss of monoacetylation and trimethylation of histone H4 is a common hallmark of human tumor cells (Fraga et al. 2005).

7.4 Histone Deacetylase Inhibitors as Epigenetic Anticancer Drugs

Cancer cells have aberrant gene expression, which has multiple causes: genetic (gene mutations, homozygous deletions, loss of heterozygosity, etc.), cytogenetic (monosomies, trisomies, homogenous staining regions, double minutes, etc.) and, of course, epigenetic (Ballestar and Esteller 2002). Epigenetic gene silencing occurs mainly through two mechanisms (Fig. 1). The first is dense hypermethylation of the CpG islands located in the promoter region of genes, followed by recruitment of MBDs and HDACs, causing gene hypomethylation (Ballestar and Esteller 2002). In the second mechanism, methylation of the gene regulatory region is not required and different transcriptional repressors aberrantly target the HDACs to the gene promoter, causing histone hypoacetylation (Fig. 1). The first route is the most thoroughly studied, and many tumor-suppressor genes, such as *p16INK4a*, *BRCA1*, *hMLH1*, *MGMT*, *VHL*, E-cadherin, etc., are known to undergo methylation-associated silencing in human neoplasms in this manner (Esteller 2002). This pathway has been targeted in cancer therapeutics using DNA-demethylating agents (Villar-Garea and Esteller 2003) that restore the functionality of silent genes, such as *p14ARF* and *hMLH1* (Esteller et al. 2001; Herman et al. 1998). A significant percentage of these drugs have toxic effects, but it is possible to reduce their dose since they have a synergistic effect on the restoration of tumor-suppressor gene expression used in combination with HDAC inhibitors (Cameron et al. 1999; Villar-Garea and Esteller 2003). HDACs alone are not able to restore the expression of hypermethylated tumor-suppressor genes. The second pathway, in which gene silencing by the promoter hypoacetylation of histones caused by recruitment of HDACs in the absence of DNA methylation, has been less studied. The archetypical gene silenced in this manner in human cancer is the cyclin-dependent kinase inhibitor *p21WAF1* (Archer et al. 1998; Blagosklonny et al. 2002; Gui et al. 2004) Epigenetic reactivation of *p21WAF1* by HDAC inhibitors has been reported in cancer cell lines (Archer et al. 1998; Blagosklonny et al. 2002), and the restoration of p21WAF1 gene expression by HDAC inhibitors is associated with enrichment of hyperacetylated histones at the p21WAF1 promoter (Gui et al. 2004). *p21WAF1* has many features that may classify it as a bona fide tumor-suppressor gene: *p21WAF1* knockout mice develop tumors (Martin-Caballero et al. 2001), *p21WAF1* expression is lost in a broad spectrum of tumor types (Shoji et al. 2002; Pellikainen et al. 2003), and its overexpression in deficient cancer cells can cause growth arrest (Archer et al. 1998). However, the lack of any gene mutation or CpG island promoter methylation may cause some concern to classical researchers: the epigenetic inactivation of *p21WAF1* by promoter hypoacetylation-mediated silencing offers an attractive alternative. Other genes altered in this manner in cancer cells continue to be identified, but more work is needed in this area.

The biochemical structure and mechanism of inhibition of HDAC inhibitors are extremely heterogeneous, from simple compounds such as valproate, to more elaborate designs, such as MS-275 (a benzamide). Overall, HDAC manifests a wide range of activity against all HDACs. The hydroxamic acids are probably the broadest set of HDAC inhibitors. The general structure of these substances consists of a hydrophobic linker that allows the hydroxamic acid moiety to chelate the cation at the bottom of the HDAC catalytic pocket, while the bulky part of the molecule acts as a cap for the tube (Fig. 3).

HDAC inhibitors may achieve some of their antitumor effects through reactivation of these new types of dormant tumor-suppressor gene. However, many other factors may also play an important role. Treatment of cancer cell lines with HDAC inhibitors has pleiotropic effects, inducing differentiation, cell-cycle arrest and apoptosis. Chimeric oncoproteins,

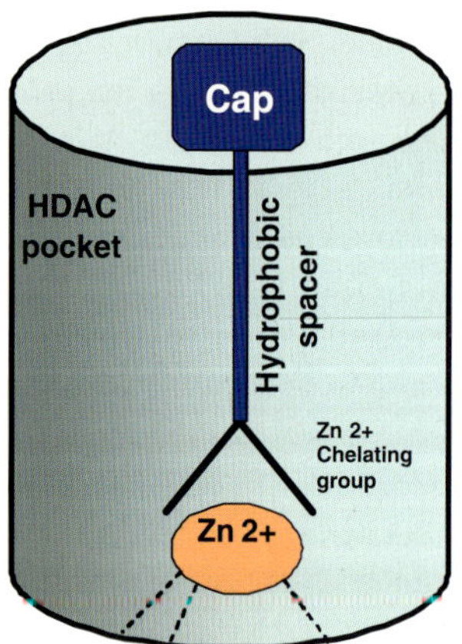

Fig. 3. Mechanism of action of HDAC inhibitors. Diagrams of the HDAC catalytic site interacting with a typical HDAC inhibitor composed by a cap, a hydrophobic spacer, and group chelating of Zn^{2+}

such as PML-RAR, PLZ-RAR and AML-ETO in hematologic malignancies, are clear examples of how differentiation is mediated by the relief of gene silencing. Nevertheless, it is possible that global genomic changes in the histone acetylation context, rather than the deacetylation of certain promoters, may be a significant step in, for example, HDAC inhibition-mediated apoptosis.

Focusing on the changes in gene expression, as we mentioned before, the presence of certain acetylated lysines in H4 and H3 tails at the promoters is generally associated with active chromatin and gene expression, so we would expect treatment of cells with HDAC inhibitors to lead to overexpression of most genes: disruption of the balance between the activity of HATs and HDACs would cause general hyperacetylation of histone tails. However, several studies of cell line growth in the presence of these drugs have revealed that the number of significantly repressed genes is similar to the number of upregulated transcripts, which is usually 4%–10%, depending on the technique used (microarray or differential display). This arises because transcription and chromatin structure are closely controlled by several different mechanisms (including other histone modifications, DNA methylation, binding of complexes and proteins, etc.), not only histone-tail acetylation, and because acetylation at certain residues may not be involved in transcription activation or may even correlate with gene silencing. HDAC inhibitors act on histone acetylation during the first 24 h after addition of the drug to the cell culture, usually peaking within the first 16 h (the exact time depends on the cell line, the compound, etc.). This effect may be reversed by growing the cells in fresh medium with no drug. Changes in the gene expression profile also appear during the first 24 h of treatment, when histones become hyperacetylated. All the known HDAC inhibitors cause an increase at both the mRNA and protein levels of the cyclin-dependent kinase inhibitor p21WAF1 as well as cell-cycle arrest. Although they act by similar mechanisms (e.g., by blocking access of acetyl-lysine residues to the HDAC catalytic pocket, which is highly conserved in all HDACs), these substances have markedly different effects: diverse patterns of gene expression are found when observing the action of several compounds in the same cell line and cell-cycle arrest can be produced at G2/M (most frequently) or at G1/S. This behavior may be due to the nonspecificity of the drug (e.g., sodium valproate also blocks ion channels) or to different affinities for the various HDACs. Also, in many systems, these chemicals can induce apoptosis through mechanisms that remain unclear but appear to be specific for each cell line (presence/absence of certain genes involved in apoptosis pathways) and drug, while in other cases these substances cause cancer cells to differentiate. HDAC inhibitors can also sensitize cancer cells to chemotherapy, and the possibility of using them in combination therapies is the most likely to succeed. For instance, cell death after treatment with etoposide, camptothecin and other substances that cross-link DNA and Topo II enzymes increases if the cell lines are pretreated with either TSA or SAHA, probably because the chromatin changes caused by the hydroxamic acids facilitate cross-linker access to the target. In other cases, HDAC inhibitors induce overexpression of a certain protein or proteins needed for the activity of another compound, such as retinoic acid receptors in the case of ATRA53 or IL-6 cytokine receptors. Some combinations have synergistic effects but their mechanism is not clear, as in the case of 1-,25-dihydroxyvitamin D3. However, the most widely reported is the synergism between demethylating agents and HDAC inhibitors that we have mentioned. The initial observation was made in colon carcinoma cell lines for hypermethylated tumor-suppressor genes such as *p16INK4a*. These drug combinations may have two advantages: first, the dose of each substance necessary for cell growth inhibition or apoptosis is usually much lower than if used separately, reducing side effects and toxicity, and second, resistance to certain chemicals can be overcome in some cases by combining drugs. The side effects are still poorly understood. HDAC inhibitors may enhance chromosomal instability. Prolonged TSA treatment alters the nuclear localization of HP1 and affects pericentromeric chromatin, causing the appearance of defective centromeres and abnormal chromosomal segregation. However, these alterations appear to be reversible, so cells recover normal cell-cycle progression and mitosis. It has also been proposed that treatment of some cancer cell lines with HDAC inhibitors, such as TSA, increases hTERT expression and may thereby contribute to tumor development, whereas in other systems, such as prostate cancer cells treated with TSA and butanoate and small-cell lung cancer cells treated with FK228, HDAC inhibitors suppress hTERT expression. Mechanisms explaining these two apparently contradictory facts have been proposed, but no definitive conclusion has been drawn.

References

Archer SY, Meng S, Shei A, Hodin RA (1998) p21WAF1 is required for butyrate-mediated growth inhibition of human colon cancer cells. Proc Natl Acad Sci U S A 95:6791–6796

Ballestar E, Esteller M (2002) The impact of chromatin in human cancer: linking DNA methylation to gene silencing. Carcinogenesis 23:1103–1109

Ballestar E, Paz MF, Valle L, Wei S, Fraga MF, Espada J, Cigudosa JC, Huang TH, Esteller M (2003) Methyl-CpG binding proteins identify novel sites of epigenetic inactivation in human cancer. EMBO J 22:6335–6345

Blagosklonny MV, Robey R, Sackett DL, Du L, Traganos F, Darzynkiewicz Z, Fojo T, Bates SE (2002) Histone deacetylase inhibitors all induce p21 but differentially cause tubulin acetylation, mitotic arrest, and cytotoxicity. Mol Cancer Ther 1:937–941

Cameron EE, Bachman KE, Myohanen S, Herman JG, Baylin SB (1999) Synergy of demethylation and histone deacetylase inhibition in the re-expression of genes silenced in cancer. Nat Genet 21:103–107

Esteller M (2002) CpG island hypermethylation and tumor suppressor genes: a booming present, a brighter future. Oncogene 21:5427–5740

Esteller M, Cordon-Cardo C, Corn PG, Meltzer SJ, Pohar KS, Watkins DN, Capella G, Peinado MA, Matias-Guiu X, Prat J, Baylin SB, Herman JG (2001) p14ARF silencing by promoter hypermethylation mediates abnormal intracellular localization of MDM2. Cancer Res 61:2816–221

Feinberg AP, Tycko B (2004) The history of cancer epigenetics. Nat Rev Cancer 4:143–153

Fraga MF, Ballestar E, Villar-Garea A, Boix-Chornet M, Espada J, Schotta G, Bonaldi T, Haydon C, Ropero S, Petrie K, Iyer NG, Perez-Rosado A, Calvo E, Lopez JA, Cano A, Calasanz MJ, Colomer D, Piris MA, Ahn N, Imhof A, Caldas C, Jenuwein T, Esteller M (2005) Loss of acetylation at Lys16 and trimethylation at Lys20 of histone H4 is a common hallmark of human cancer. Nat Genet 37:391–400

Gui CY, Ngo L, Xu WS, Richon VM, Marks PA (2004) Histone deacetylase (HDAC) inhibitor activation of p21WAF1 involves changes in promoter-associated proteins, including HDAC1. Proc Natl Acad Sci U S A 101:1241–1246

Herman JG, Umar A, Polyak K, Graff JR, Ahuja N, Issa JP, Markowitz S, Willson JK, Hamilton SR, Kinzler KW, Kane MF, Kolodner RD et al. (1998) Incidence and functional consequences of hMLH1 promoter hypermethylation in colorectal carcinoma. Proc Natl Acad Sci U S A 95:6870–6875

Iizuka M, Smith MM (2003) Functional consequences of histone modifications. Curr Opin Genet Dev 13:154–160

Jenuwein T, Allis CD (2001) Translating the histone code. Science 293:1074–1080

Johnstone RW (2002) Histone-deacetylase inhibitors: novel drugs for the treatment of cancer. Nat Rev Drug Discov 1:287–299

Jones PA, Baylin SB (2002) The fundamental role of epigenetic events in cancer. Nat Rev Genet 3:415–428

Martens JH, O'Sullivan RJ, Braunschweig U, Opravil S, Radolf M, Steinlein P, Jenuwein T (2005) The profile of repeat-associated histone lysine methylation states in the mouse epigenome. EMBO J 24:800–812

Martin-Caballero J, Flores JM, Garcia-Palencia P, Serrano M (2001) Tumor susceptibility of p21Waf1/Cip1-deficient mice. Cancer Res 61:6234–6238

Pellikainen MJ, Pekola TT, Ropponen KM, Kataja VV, Kellokoski JK, Eskelinen MJ, Kosma VM (2003) p21WAF1 expression in invasive breast cancer and its association with p53, AP-2, cell proliferation, and prognosis. J Clin Pathol 56:214–220

Peterson CL, Laniel MA (2004) Histones and histone modifications. Curr Biol 14:R546–R551

Shoji T, Tanaka F, Takata T, Yanagihara K, Otake Y, Hanaoka N, Miyahara R, Nakagawa T, Kawano Y, Ishikawa S, Katakura H, Wada H (2002) Clinical significance of p21 expression in non-small-cell lung cancer. J Clin Oncol 20:3865–3871

Strahl BD, Allis CD (2000) The language of covalent histone modifications. Nature 403:41–45

Turner BM (2000) Histone acetylation and an epigenetic code. Bioessays 22:836–845

Turner BM (2005) Reading signals on the nucleosome with a new nomenclature for modified histones. Nat Struct Mol Biol 12:110–112

Villar-Garea A, Esteller M (2003) DNA demethylating agents and chromatin-remodelling drugs: which, how and why? Curr Drug Metab 4:11–31

8 Histone Post-Translational Modifications Regulate Transcription and Silent Chromatin in *Saccharomyces cerevisiae*

N.C. Tolga Emre, S.L. Berger

Abstract. Regulation of chromatin structure is important for the control of DNA-templated processes such as gene expression and silencing, and its dysregulation is implicated in diverse developmental and cell proliferative defects such as tumorigenesis. Covalent post-translational modifications of histones are one of the prominent means to regulate the chromatin structure. Here, we summarize findings from our lab and others regarding the interactions between different covalent modifications of histones in the budding yeast *Saccharomyces cerevisiae.* First, we describe the effect of histone H3 phosphorylation at residue serine 10 in transcriptional gene activation, and its histone H3 acetylation dependent and independent modes of action and downstream effects on TATA-binding protein (TBP) recruitment. Further, we review how ubiquitylation of histone H2B and its deubiquitylation by ubiquitin proteases Ubp8 and Ubp10 regulate histone H3 methylations, and consequently affect co-activator-dependent gene transcription and silent chromatin, respectively.

8.1 Introduction

The basic unit of chromatin, the nucleosome core particle, consists of approximately 200 base pairs of DNA wrapped around an octamer made up of core histones H2A, H2B, H3, and H4. The successive nucleosomes connected by free stretches of linker DNA and/or linker histone H1 constitute the primary chromatin structure, termed "bead on a string" (Khorasanizadeh 2004; Luger 2003). In situ, higher-order chromatin folding, such as 30-nm fibers to condensed metaphase chromosomes is also observed, which provides an increasing packing ratio for DNA. This ordered organization of genomic DNA in chromatin provides a solution to space constraints imposed on the genome by the eukaryotic nucleus. However, DNA-dependent molecular processes such as gene transcription, DNA repair, and recombination are generally hindered by this compaction. The studies on RNA polymerase II-dependent gene transcription has historically been focused more on the basal RNA polymerase II machinery and DNA-bound factors such as activators and repressors. This development has been against a backdrop of early studies linking certain histone modifications (e.g., acetylation) with certain transcriptional states (e.g., activation). The discovery that activator-associated factors (coactivators and corepressors) in transcriptional regulation act as histone-modifying activities (Brownell et al. 1996; Hassig et al. 1997) led to a surge of interest in histone modifications in genomic regulation, starting mostly with gene transcription. Modulation of the chromatin structure not only provides a means to overcome the general hindrance of the nucleosome structure on DNA, but also provides added regulatory opportunities for diverse nuclear processes. ATP-dependent chromatin remodeling activity, use of histone variants, and covalent post-translational modifications of histones all provide means to modulate the chromatin structure (Vaquero et al. 2003).

8.1.1 Major Covalent Modifications of Histones: A Brief Overview

Early biochemical studies have pointed out a variety of post-translational modifications on histones (for example, Allfrey et al. 1964). Today we know that histones are post-translationally modified on a variety of conserved residues with the addition of phosphoryl, acetyl, methyl and ubiquityl groups (Berger 2002), among others. In many instances, these modifications have been correlated with distinct chromosomal and transcriptional states. In this regard, acetylation (ac) of core histone tails has long been correlated with transcriptionally active loci (Wolffe 1998). Several residues have been identified as acetylated on all major histones (reviewed in Sterner and Berger 2000), and this list has grown even longer with the recent implementation of proteomics techniques on histones (Cosgrove et al. 2004; Freitas et al. 2004). The best-studied acetylation sites in budding yeast are H3K9ac and H3K14ac in gene activation and H4K16ac implicated in gene silencing (reviewed in Kurdistani and Grunstein 2003).

A major leap in understanding the role of histone acetylation came when a *Tetrahymena* homolog of Gcn5 was shown to possess intrinsic histone acetyltransferase (HAT) activity (Brownell et al. 1996). In this study, Gcn5, a previously known transcriptional coactivator (Marcus et al. 1994), was shown to acetylate histone tails as a part of SAGA (Spt-Ada-Gcn5 acetyltransferase) multiprotein complex (Kuo and Allis 1998). Further studies demonstrated that HAT activity is necessary for Gcn5's role as a coactivator (Candau et al. 1997; Grant et al. 1997). Several other previously known coactivators in diverse organisms such as in budding yeast (e.g., Esa1) and in humans (e.g., CBP/p300) were then shown to contain HAT activity in vitro and in vivo, and their HAT activity is necessary for maximal transactivation of their target genes (reviewed in Sterner and Berger 2000). Complementary studies on transcriptional corepressors (such as Sin3p of *S. cerevisiae*, and mammalian mSin3) revealed that corepressors function by recruiting histone deacetylases (HDACs) to the promoters (Struhl 1998).

Phosphorylation (ph) is one of the common modifications on histones, observed on all the core histones and the linker histone H1, on multiple serine residues. Threonine residue phosphorylation is also observed, but our understanding is still limited (Freitas et al. 2004; Polioudaki et al. 2004). Like acetylation, phosphorylation of histones has also been implicated in various processes such as gene transcription and DNA repair; however, unlike acetylation, histone phosphorylation can mediate seemingly contradictory outcomes, even at a given particular residue: phosphorylation of serine 10 on histone H3 N-terminal domain (H3S10ph, as suggested in Turner 2005) is very well correlated with mitotic condensation of chromosomes and provides a good marker for mitosis (reviewed in Nowak and Corces 2004). Condensed chromosome structure in mitosis is generally repressive for gene transcription, especially in higher eukaryotes. On the other hand, the same phosphorylation mark on H3 Ser10 (together with histone H3 acetylation in some cases) is also associated with immediate early gene expression in mammalian cells (Thomson et al. 1999 and references therein), NF-KB induction (Anest et al. 2003; Yamamoto et al. 2003) and with activated gene transcription in budding yeast (Lo et al. 2000, 2001) (see Sect. 8.2 for details).

Methylation of histones occurs on various arginine and lysine residues and it is one of the historically earliest detected modifications. In contrast to acetylation and phosphorylation, where the target residue can be modified by only a single moiety, each arginine can be modified by up to two methyl groups (mono- or dimethyl), and lysines can be modified up to three methyl groups (i.e., trimethyl lysine). These distinct methylation levels of a given residue have been linked to distinct genomic activities in some cases, but the significance is unknown for others (reviewed in D.Y. Lee et al. 2004). For instance, while all the euchromatic regions tested show dimethylation of H3 Lys4 (H3K4me2) in budding yeast, trimethylation of Lys4 (H3K4me3) preferentially occurs at promoters and 5′ regions of active ORFs (Bernstein et al. 2002; Santos-Rosa et al. 2002), likely due to untargeted and targeted activity of the corresponding histone methyl transferase, Set1, respectively (Ng et al. 2003b). A similar spatial confinement among different levels of methylated H3 Lys9 residue has been observed for different classes of heterochromatin in mammalian cells (for example, Rice et al. 2003). How differential levels of methylation of a given residue may contribute to different functional outcomes (if at all) remains largely unknown (reviewed in Sims et al. 2003). Methylation also plays a dual role in gene transcription: for example, H3 Lys4 and Lys79 methylations (H3K4me, H3K79me) are associated with active chromatin in all the organisms tested (including budding yeast, where they affect silencing indirectly, see Sect. 8.4). On the other hand, H3 Lys9 and Lys27 methylations (H3K9me, H3K27me) are required for heterochromatic silencing and/or gene-specific repression in many eukaryotes, by virtue of their binding to chromodomains of HP1 and polycomb (Pc) class of proteins, respectively (reviewed in Sims et al. 2003). These latter modifications and their chromodomain-containing binding partners have not been detected in the budding yeast *S. cerevisiae*.

Histone methylation is considered as a prime candidate among chromatin modifications for an epigenetic cellular memory mark that keeps track of cell identity across mitotic divisions (Turner 2002). The reasons for this are several: First, the methylation status of certain histone residues has been shown to be relatively stable across generations (for example, see Ng et al. 2003b). Secondly, particular methylations, such as H3K9me and H3K27me, are required for the maintenance of heterochromatic states in a broad range of organisms. Furthermore, methylation, especially on lysines, is a thermodynamically stable modification, and no histone demethylating enzymes have been identified until very recently. The recent demonstration that a mammalian nuclear FAD-dependent amine oxidase enzyme homolog (named LSD1, for lysine specific demethylase 1) reverses specifically the Lys4 di- and monomethylation of H3 in vitro and in vivo by a two-step amine oxidation reaction (Shi et al. 2004) cast doubt on the latter reasoning (Kubicek and Jenuwein 2004). Nevertheless, currently there is no evidence that argues against a cellular memory role for histone lysine methylation because of demethylating enzymes, since the only known demethylase LSD1 is highly specific for H3 meLys4 only, but not other residues. Furthermore, the enzymatic pathway used by LSD1 cannot act on trimethylated lysines (Shi et al. 2004), which is thought to be the major "memory mark" in most of the cases (D.Y. Lee et al. 2004; Turner 2002).

The 76-residue conserved polypeptide ubiquitin (ub) is covalently attached to many cellular proteins. Multiple rounds of ubiquitylation of a given lysine residue, leading to polyubiquitin chains, was first shown to result in proteosome-mediated degradation of the target protein. Ubiquitin conjugation mediated by the sequential action of E1-activating enzymes and E2-conjugating enzymes works with E3 ligases, which transfer ubiquitin to their cognate substrates via an isopeptide bond involving its C-terminal glycine residue (reviewed in Hochstrasser 1996; Wilkinson 2000). In addition to its role in protein turnover, regulatory and signaling roles of attachment of a single ubiquitin molecule (i.e., "monoubiquitylation") to a variety of cellular proteins have been uncovered in recent years (Aguilar and Wendland 2003; Hicke 2001).

Ubiquitylated forms of histones were first observed more than two decades ago, and changes in the ubiquitylation status of individual histones have been correlated with distinct cellular states and processes, such as transcriptional activation (reviewed in Osley 2004; Zhang 2003). However, molecular mechanisms involved in the ubiquitylation of histones and the universality of their functional consequences has remained relatively less well-defined over the years. More recently, it has been shown that budding yeast histone H2B is monoubiquitylated by the ubiquitin ligase Rad6 in vivo at the residue Lys123 (H2BK123ub1) and an inability to ubiquitylate causes distinct mitotic and meiotic cell division defects (Robzyk et al. 2000). Later studies have identified Bre1 as the ubiquitin ligase that directs Rad6-dependent ubiquitylation activity to histone H2B (Hwang et al. 2003; Wood et al. 2003).

8.1.2 Cross-talk Among Histone Modifications and Their Functional Correlates

How do covalent modifications of histones affect DNA-templated processes? Two prevailing hypotheses (and their variants) exist, although they need not be mutually exclusive and their precedence may be context-dependent. The first hypothesis, mainly derived from the fact that acetylation changes the overall charge on histones, posits that modifications can disrupt nucleosome and/or higher-order chromatin structure due to their neutralizing effect on the overall charge of histones or by their sheer existence, thus permitting protein factors (such as transcription factors) to reach their DNA targets. This hypothesis is currently backed more by in vitro studies (reviewed in Elgin and Workman 2000), but some in vivo evidence also exists (for example, Ren and Gorovsky 2001). The second hypothesis focuses on accumulating evidence that covalently modified residues can specifically bind certain protein domains, suggesting that modifications exist to recruit effector proteins to the chromatin for specific tasks. For instance, bromodomains can specifically bind acetylated lysine residues, while chromodomains bind methylated lysines, and SANT domains preferentially bind unmodified histone tails (for specific examples and details, please refer to recent reviews: Bottomley 2004; Boyer et al. 2004; de la Cruz et al. 2005).

A substantial portion of the proteins with motifs that bind covalently modified histones are themselves chromatin-modifying enzymes or part of such complexes. This leads to cross-talk between histone modifications, where the presence (or absence) of a given histone modification facilitates (or inhibits) another modification (reviewed in Fischle et al. 2003). Just to mention a few examples among many, Set1-dependent methylation of H3 tail at Lys4 promotes the association of Iswi1 ATPase during gene activation in budding yeast (Santos-Rosa et al. 2003), while the same modification by Set9 in human cells prevents the binding of HDAC-containing corepressor NuRD with chromatin (Nishioka et al. 2002).

An intriguing aspect of histone H2B ubiquitylation in budding yeast is its intimate connection with H3K4me and H3K79me. This is the only known trans-tail cross-talk between histone modifications, where H3K4me and H3K79me require ubiquitylation of H2B Lys123, but not vice versa (Briggs et al. 2002; Dover et al. 2002; Sun and Allis 2002). Although the mechanisms underlying the trans-tail dependence are not yet understood, this cross-talk is likely the main link that connects histone H2B ubiquitylation to various processes such as gene transcription and heterochromatin-like silencing, as we discuss in Sect. 8.4.

The budding yeast *S. cerevisiae* has been a favorite eukaryotic model system for the investigation of various cellular processes as diverse as cytoskeletal dynamics and cell cycle control (Botstein and Fink 1988). This list also extends to gene transcription and chromatin function. Among many properties of budding yeast that make it a suitable model organism is the high conservation of basic processes with higher eukaryotes. A majority of the histone modifications and the enzymes (and their complexes) that regulate these modifications are conserved among budding yeast and mammalian systems, and the initial insights into their functions and mechanisms usually come from studies on budding yeast.

In the next section, we summarize the recent findings from our laboratory and related studies by others pertaining to the mechanisms and consequences of cross-talk among different histone modifications.

8.2 Histone H3 Ser10 Phosphorylation and Acetylation Regulate TATA Binding Protein Association During Transcriptional Initiation

We and others have previously examined the role of and cross-talk between H3K14ac and H3S10ph in activated gene transcription. These studies uncovered a synergy among the two modifications in activation of a subset of SAGA (Gcn5)-dependent genes in budding yeast (Lo et al. 2000, 2001), and growth-factor stimulated immediate early gene *c-fos* in mammalian cells (Mahadevan et al. 1991). Phosphoacetylation of histone H3 during *c-fos* induction has been well documented (reviewed in Thomson et al. 1999). In vitro enzyme assays with Gcn5 enzyme and SAGA complex on synthetic H3 peptides revealed that Ser10 phosphorylated peptides are better substrates for acetylation, providing a mechanistic link for the observed synergy among these modifications (Cheung et al. 2000; Lo et al. 2000). Additionally, we have identified Snf1 (Sucrose nonfermenting 1) as the histone H3 Ser10 kinase during the induction of the *INO1* gene in response to inositol starvation in budding yeast, and showed that Snf1-dependent Ser10 phosphorylation and Gcn5-dependent H3K14ac occur in a temporal order during induction (Lo et al. 2001). Further studies using Gcn5 and H3 tail peptides provided a structural view for the observed binding preference of Gcn5 to H3S10ph compared to the nonphosphorylated H3 peptide (Clements et al. 2003). In other contexts, such as the heat shock response in *Drosophila*, H3S10ph during gene activation is not linked to histone acetylation, suggesting that coupling of these modifications is probably promoter-dependent (reviewed in Nowak and Corces 2004).

Yet another level of versatility in the modulation of promoter activity by H3S10ph is demonstrated among SAGA-dependent, inducible genes in budding yeast. We found that Snf1-dependent H3S10ph is also important for normal *GAL1* induction (similar to what has been observed for *INO1*). SAGA recruitment and H3K14ac at the core promoter proceeds that of H3S10ph during the induction time course at *INO1*. Similarly, the Snf1 kinase complex interacts with transcriptional activators of both *INO1* (i.e., Ino2) and *GAL1* (i.e., Gal4) in vitro. However, distinct mechanisms prevail for SAGA recruitment: Gal4 activator interacts with SAGA (in line with previous observations; Bhaumik and Green 2001;

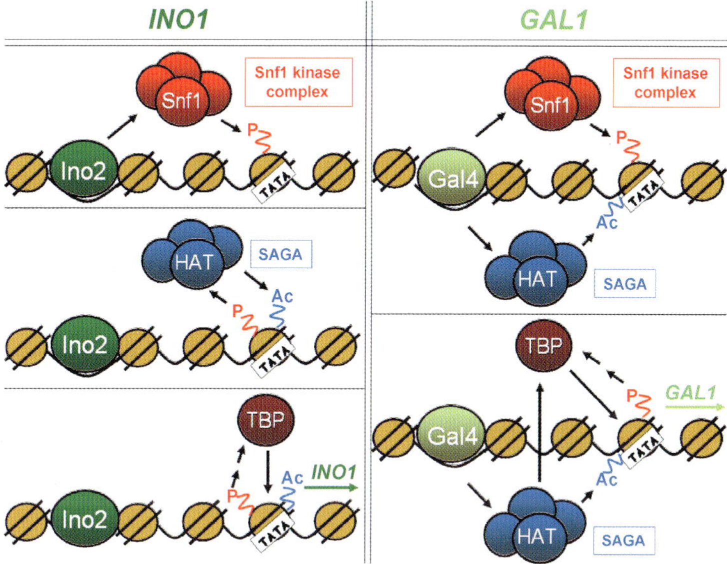

Fig. 1. A model for common and distinct roles of H3S10ph at two SAGA-regulated genes in the budding yeast. At *INO1* locus (*left panel*), Ino2-dependent recruitment of Snf1 complex sets up the H3S10ph mark that is instrumental in SAGA recruitment. At the *GAL1* locus (*right panel*), Snf1 and SAGA complex recruitments apparently occur independently, both through the activator Gal4. In either case, H3S10ph mark is involved in TBP association with the TATA box. A role for SAGA complex has also been implicated in TBP recruitment to the *GAL1* locus (*right panel*). Only relevant factors depicted, not to scale

Larschan and Winston 2001) while Ino2 does not. Additionally, while at the *INO1* promoter, both SAGA recruitment and H3K14ac are defective in the absence of Snf1 or H3S10ph, but they are not affected at the *GAL1* promoter (Lo et al. 2005). Thus, given the preferential binding of Gcn5 to the Ser10 phosphorylated tail as described above (Cheung et al. 2000; Clements et al. 2003; Lo et al. 2000), it is possible that SAGA complex requires interaction with the phosphorylated H3 tail for efficient recruitment to the *INO1* promoter, since it cannot be recruited by the Ino2 activator through direct interaction. On the other hand, Gal4 can directly interact with and recruit SAGA to *GAL1* (Bhaumik and Green 2001; Larschan and Winston 2001; Lo et al. 2005), which overrides the need for an interaction with the phosphorylated tail, thus leading to a lack of dependence of H3K14ac on prior H3S10ph. Despite this difference in SAGA recruitment, both promoters share a common downstream effect: binding of TATA-binding protein (TBP) to its cognate site on DNA (which is important for preinitiation complex formation) at any of these promoters upon induction is compromised in the absence of H3S10ph (Lo et al. 2005) (Fig. 1).

8.3 Histone H2B Ubiquitylation and Deubiquitylation Regulate Histone H3 Methylation During Transcription

One of the earliest indications of involvement of histone H2B ubiquitylation in transcriptional gene regulation came from the observation that H2B ubiquitylation mutants genetically interact with the Gcn5 HAT component of the SAGA complex during induced expression of certain budding yeast genes, such as *GAL1* (Kao et al. 2004; Osley 2004). Further studies showed that, indeed, the histone H2B ubiquitin conjugase Rad6 is recruited to the promoter of *GAL1* during gene induction in a transcriptional activator- and Bre1-dependent manner. This recruitment coincides with increased H2B ubiquitylation and SAGA recruitment at the promoter-proximal regions; however, the recruitment of Rad6 and SAGA are independent (Kao et al. 2004).

Similar to the reversibility of other post-translational modifications by specific enzymes, ubiquitin conjugation is also countered by the action of specific protease enzymes, collectively called deubiquitylating enzymes

(DUBs) (reviewed in Soboleva and Baker 2004; Wilkinson 1997, 2000). The largest family of DUBs are the ubiquitin proteases (abbreviated as UBP in budding yeast; USP in mammalians). The budding yeast genome contains 16 such putative thiol-proteases as judged by signature catalytic motifs called the cys- and his-box, which are required for the formation of a catalytic triad to cleave the isopeptide bond between the substrate and the ubiquitin, or between two ubiquitin molecules in a polyubiquitin chain. Although DUBs are implicated in diverse biological processes in many organisms, the substrate specificity and regulation of only a few has been characterized (reviewed in Soboleva and Baker 2004). In this regard, no specific in vivo deubiquitylating activity toward histones has been characterized until recently.

A large-scale proteomic study has identified the putative deubiquitylating enzyme Ubp8 as a novel subunit of the coregulator complex SAGA (Sanders et al. 2002). This finding led us to hypothesize that Ubp8 targets ubiquitylated histone H2B, given the genetic interactions observed between the SAGA complex and histone H2B ubiquitylation, as explained above. Consequently, we and others found that Ubp8 is a stable and stoichiometric subunit of SAGA with a role in H2B deubiquitylation in vitro and in vivo (Daniel et al. 2004; Henry et al. 2003). Ubp8 is recruited to the SAGA-dependent *GAL1* gene promoter upon induction with galactose (Daniel et al. 2004; Henry et al. 2003). This recruitment is coincident with SAGA complex recruitment, and leads to decreased H2BK123ub1 levels (Henry et al. 2003). Strikingly, we also found that both ubiquitylation and deubiquitylation of H2B are required for the proper activation of certain SAGA-dependent genes (Henry et al. 2003), a finding that is contrary to what is typically observed for the antagonistic functional outcomes of reversible covalent modifications (for example, acetylation of histones at promoters usually associated with gene activation, while deacetylation of the same residues with repression). We studied the basis of this unusual requirement and found that the lack of H2B ubiquitylation exhibits different defects in H3 methylation compared to those in the absence of H2B deubiquitylation. Targeted H3K4me3 occurs at the 5′ end of the ORF early during transcript elongation (Ng et al. 2003b; Santos-Rosa et al. 2002), while H3K36me occurs primarily later during transcript elongation (Krogan et al. 2003; Xiao et al. 2003). We found that histone H2B ubiquitylation at *GAL1* promoter is required for H3K4me (in line with previous studies: Briggs et al. 2002; Dover et al. 2002; Sun and Allis 2002), while deubiquitylation is required for H3K36me, leading to the proposal that sequential ubiquitylation and deubiquitylation of H2B establishes the correct balance among different H3 methylations required for proper gene activation (Henry et al. 2003; Wyce et al. 2004) (Fig. 2).

Except for the deubiquitylating enzymes that are part of the proteasome, little is known about how they are targeted to substrates for specificity (Soboleva and Baker 2004). In the case of Ubp8, integration of Ubp8 SAGA multiprotein complex likely delivers Ubp8 to the appropriate position at the proper time to provide substrate specificity. However, it was unknown how Ubp8 is associated into SAGA. We investigated the basis for Ubp8 association with SAGA and noted an amino terminal cysteine/histidine rich region. This region is predicted to form a zinc binding module (i.e., a zinc finger, or ZnF) called ZnF_UBP (Pfam domain ID#: PF02148; Bateman et al. 2004). A similar ZnF_UBP domain is also present in murine HDAC6 (a histone deacetylase), where it is involved in ubiquitin binding (Hook et al. 2002; Seigneurin-Berny et al. 2001), and is also found in other deubiquitylating enzymes. Using a panel of deletion and substitution mutants in the ZnF of Ubp8, we found that it is important for association of Ubp8 with SAGA and for Ubp8's deubiquitylating activity in vivo. On the other hand, catalytic domain mutations of Ubp8 reduce its enzymatic activity on ubiquity-

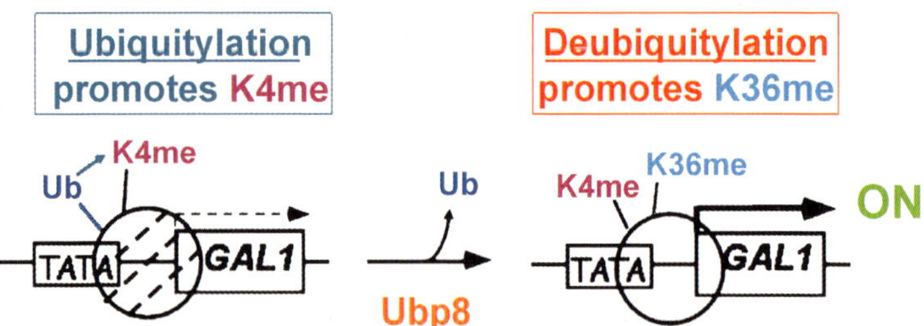

Fig. 2. A role for Ubp8-dependent histone H2B deubiquitylation at the *GAL1* locus. Ubiquitylation of histone H2B is required for subsequent H3K4me (see text for details). Deubiquitylation by Ubp8 of the same residue is implicated in proper H3K36me levels at the *GAL1* promoter and transcriptional activation

lated H2B as expected, but not association with SAGA (Ingvarsdottir et al. 2005). Furthermore, either catalytic or zinc finger mutants are defective in pathways requiring SAGA when combined with deletion of *GCN5*, suggesting that Ubp8 exerts its effects through histone H2B deubiquitylation and as a part of SAGA complex (Ingvarsdottir et al. 2005).

We and others recently identified Sgf11 (*SAGA* associated *f*actor 11 kD) as another novel component of SAGA (Ingvarsdottir et al. 2005; K.K. Lee et al. 2004; Powell et al. 2004) and showed that Sgf11 is important for the association of Ubp8 with SAGA (Ingvarsdottir et al. 2005; Powell et al. 2004) and for Ubp8's deubiquitylation activity on histone H2B (Ingvarsdottir et al. 2005). Ubp8 interacts with Sgf11 in vivo and the ZnF domain of Ubp8 is required for this interaction (Ingvarsdottir et al. 2005). We also showed that the functions of Ubp8 and Sgf11 are related and separable from other components of SAGA, such as the TBP regulatory module (composed of Spt3/Spt7/Spt8) and the histone acetyltransferase module (composed of Gcn5/Ada2/Ada3). In this regard, the profiles of *UBP8* and *SGF11* deletions are similar to each other in global gene expression analyses and large-scale synthetic genetic interactions (Ingvarsdottir et al. 2005). Based on these observations, we suggest that Ubp8 and Sgf11 represent a new functional module within the SAGA complex (termed the histone deubiquitylase [HDUB] module) involved in gene regulation through H2B deubiquitylation, which is consistent with the modular construction of this complex (Grant et al. 1998). Whether Ubp8 and Sgf11 represent a separable, physical module is yet to be established.

8.4 Histone H2B Deubiquitylation Regulates Heterochromatic Silencing

8.4.1 Silent Chromatin in Budding Yeast

In contrast to activator-mediated local recruitment of chromatin modification complexes to regulate transcription, other DNA-templated processes require broader recruitment, either to certain regions of the genome, or even to whole chromosomes. We have identified a second

H2B deubiquitylating enzyme, Ubp10, that helps to maintain quiescence of certain regions within the budding yeast genome (Emre et al. 2005).

In budding yeast, the cytological dense-staining of heterochromatin cannot be observed, probably due to the relatively open structure of the budding yeast chromatin. However, certain regions on budding yeast chromatin show strong functional correlates to metazoan heterochromatin, namely late replication and repression of DNA-templated processes, such as gene transcription (Perrod and Gasser 2003 and references therein). Heterochromatin-like silencing in budding yeast is formed through the action of certain DNA-binding factors (such as Rap1, Abf1, Orc1), the silent information regulatory (SIR) complex, and an indirect but crucial contribution from certain histone modifications such as acetylation, methylation, and ubiquitylation (see Sect. 8.4.2). Silenced regions in budding yeast include telomere-proximal regions (i.e., 5–20 kb from the ends of the chromosomes), silent mating-type loci (*HMR* and *HML*) where nonexpressed copies of sex (mating type)-specific gene reside, and the rDNA locus where tandem copies of ribosomal RNA genes exist (reviewed in Huang 2002; Rusche et al. 2003). Sir2, an NAD^+-dependent HDAC, is required for the silencing of all three types of loci mentioned above. Budding yeast Sir2 can deacetylate histones in vitro and in vivo, and its main target is the Lys16 residue of H4.

RNAP II-dependent gene transcription is repressed in telomere-proximal, Sir-repressed regions of the chromosomes in a majority of the cells, a phenomenon called telomere position effect (TPE). TPE is inherited epigenetically in a clonal fashion, with low rates of reversion, akin to position effect variegation (PEV) in other eukaryotes (Gottschling et al. 1990). A similar TPE is also observed in mammalian cells, although underlying mechanisms may be different (Baur et al. 2001).

Silent chromatin at chromosome ends is thought to contribute to genomic stability since, if left in an “open” confirmation, these ends resemble double-strand breaks and are prone to DNA repair and recombination. Sir proteins are initially recruited by DNA-binding proteins such as Rap1, which specifically bind DNA elements on the telomeric repeats or on the region called silencers that flank the silent mating-type loci. After this nucleation event, the Sir complex spreads into the chromosome, or over the mating type loci, by virtue of Sir3’s ability to interact with the histone tails that are underacetylated and undermethylated (see the next section), leading to a chromatin structure that is refractory to DNA-based processes (Gasser and Cockell 2001; Perrod and Gasser 2003).

Earlier studies demonstrated that histones proximal to telomeres are hypoacetylated (Grunstein 1997; Rusche et al. 2003). This property is thought to be both the reason and consequence of Sir complex occupancy at the telomere-proximal chromatin. Firstly, Sir complex bound chromatin may not be an efficient substrate for HATs. Secondly, Sir2 actively deacetylates histones. Finally, Sir3 binds to hypoacetylated histone tails better, as exemplified with in vitro binding assays and in strains with severe acetylation defects, where Sir proteins extend their binding targets into normally nonsilenced chromatin. Recent studies suggest spreading of Sir proteins from the nucleation points is counteracted both by acetylation through the HAT Sas2 (Kimura et al. 2002; Suka et al. 2002) and by a specialized H2A variant, H2A.Z (Meneghini et al. 2003) and other proteins (Ladurner et al. 2003) at the boundary regions between silent and active chromatin.

Observations very similar to histone acetylation have recently been made for H3K4me and H3K79me at budding yeast silent chromatin. These modifications (especially H3K4me2 and H3K79me3) are ubiquitous in the active, euchromatic regions of the genome; however, they are much reduced in the silenced domains (Bernstein et al. 2002; Briggs et al. 2001; van Leeuwen and Gottschling 2002). Further, Sir3 binds to histone H3 peptides nonmethylated at Lys4 better than the methylated form in vitro. Additionally, mutations in H3K4me and H3K79me cause Sir protein spreading into the active regions of the genome in vivo, paralleling what has been observed for histone acetylation. These observations prompted a model to explain the relationship between global histone modifications and silencing (van Leeuwen and Gottschling 2002): lack of global modifications on histone tails (such as lack of H3K4me and H3K79me, or lysine acetylation) leads to loss of Sir proteins (whose levels are limited in the cell) away from silenced regions, causing silencing defects. Therefore, although there is no apparent mark to directly promote heterochromatin in budding yeast akin to metazoan H3K9me or H3K27me, histone modification marks that identify euchromatin indirectly control silent chromatin, by regulating Sir protein partitioning across active and silent chromatin.

8.4.2 Ubiquitin Protease Ubp10 Regulates Silencing Through Histone H2B Deubiquitylation

Our identification of Ubp8 as a histone H2B-deubiquitylating enzyme prompted us to investigate whether any of other budding yeast ubiquitin proteases might be involved in histone H2B deubiquitylation in transcriptional gene regulation.

UBP10/DOT4 (*D*isruptor *o*f *T*elomeric Silencing 4) was initially isolation in a screen for high copy disruptors of telomeric silencing in budding yeast. Both deletion and overexpression of *UBP10* cause silencing defects (Kahana and Gottschling 1999; Singer et al. 1998). In addition, Ubp10 possesses deubiquitylating activity toward a generic ubiquitylated substrate and this catalytic activity is important for its pro-silencing activity. Furthermore, *UBP10* interacts with *SIR4* in a budding yeast two-hybrid assay, proving a clue in its function in silencing (Kahana and Gottschling 1999).

Recent studies have also identified a role for Ubp10 in budding yeast apoptosis. Strains lacking *UBP10* show sensitivity to H_2O_2 and display other marks of budding yeast apoptosis. A genome-wide study of *UBP10* and SIR mutants provided additional evidence for the telomeric silencing role of *UBP10* and linked the pro-apoptotic phenotype of *UBP10* mutants to telomere-proximal transcriptional regulation abnormalities in these *UBP10*-null strains (Bettiga et al. 2004; Orlandi et al. 2004). However, neither *UBP10*'s mechanism of action nor its deubiquitylation substrates in silencing have previously been identified. We focused on the mechanisms by which Ubp10 regulates telomeric gene silencing.

We found that the level of H2BK123ub1 is increased in a strain deleted for *UBP10*, similar to what is observed for *UBP8* deletion. In vitro deubiquitylation assays using purified recombinant Ubp10 and ubiquitylated H2B further confirm that H2B is a substrate for Ubp10, suggesting that H2BK123ub1 is a direct substrate of Ubp10. In addition, chromatin immunoprecipitation (ChIP) assays demonstrate that deletion of *UBP10* leads to increased H2BK123ub1 specifically at a telomere proximal region, where Ubp10 preferentially localize, also as judged by ChIP assays. Further, in the absence of Ubp10, expression of a normally silenced gene within this affected region is increased (Emre et al. 2005).

These data suggest that Ubp10 is directed to telomere-proximal regions to maintain low ubiquitylation leading to low gene expression.

We further investigated the molecular mechanisms of Ubp10 in silencing. We hypothesized that loss of Ubp10 may alter histone H3 methylation levels and Sir protein localization at the telomere, since previous studies provide evidence that H2BK123ub1 is required for H3K4me and H3K79me in euchromatic regions of the genome, and especially in ORFs, as explained Sect. 8.3. Mutations that prevent these methylations cause silencing defects, likely due to relocalization of Sir proteins from silenced regions to active regions of the genome (van Leeuwen and Gottschling 2002). Accordingly, we tested whether the increase in H2BK123ub1 in the absence of Ubp10 leads to an increase in methylated H3 forms. We found that overexpression of Ubp10 leads to loss of global H2BK123ub1 and loss of H3K4me and H3K79me as well. Furthermore, loss of Ubp10 also causes increased H3K4me and H3K79me close to the telomere but not at active regions (Emre et al. 2005). Interestingly, loss of Ubp10 and loss of Sir2 mutually lower each other's association at the telomere. Further, ChIP analysis demonstrates Ubp10 association with rDNA regions where Ubp10 deletion also leads to increased H3 methylation (Emre et al. 2005). In a recent study, similar observations were made regarding the localization of Ubp10 to silent chromatin and its role in the regulation of histone H2B ubiquitylation, H3 methylation and telomere-proximal gene expression (Gardner et al. 2005).

These findings suggest the following model: Ubp10 is preferentially associated with silent chromatin and deubiquitylates H2B in those regions, contributing to low methylation, and favoring Sir association. Loss of Ubp10 causes an increase in H2BK123ub1, leading, in turn, to increased methylated H3, which is unfavorable for Sir protein association with the chromatin (Ng et al. 2003a; Santos-Rosa et al. 2004; van Leeuwen and Gottschling 2002). Decreased Sir2 binding in turn leads to decreased Ubp10 association with chromatin presumably because of the role of Sir proteins in the recruitment of Ubp10. Thus there is likely a feedback loop between the HDAC and HDUB activities to maintain low modifications and proper silencing. On the other hand, overexpression of Ubp10 leads to global deubiquitylation of histone H2B (and thus global hypomethylation of H3 Lys4 and Lys79), which results in promiscuous binding of Sir proteins to normally active chromatin, causing the limited pools of Sirs to escape from silenced regions, yet another reason for disruption of silencing (Emre et al. 2005) (Fig. 3).

We have also investigated any potential overlap among the known histone H2B deubiquitylating enzymes Ubp8 and Ubp10. Our results argue for distinct roles of Ubp8 and Ubp10 in SAGA-mediated gene activation and telomeric silencing, respectively.

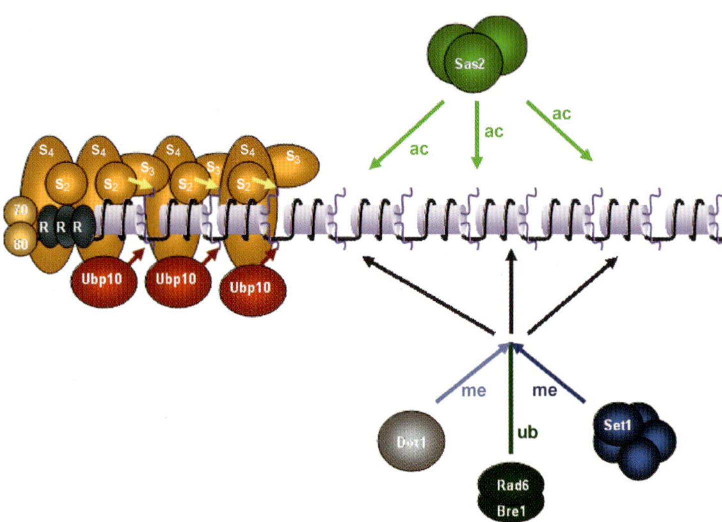

Fig. 3. A model for the involvement of histone covalent modifications in telomeric silencing in the budding yeast. Telomere proximal regions are silenced due to Sir protein occupancy (2, Sir2; 3, Sir3; 4, Sir4), initially mediated by the DNA binding proteins, such as Ku70/80 (70, 80) and Rap1 (*R*). Further spreading of the Sir complex into the chromosome is dependent on Sir2-dependent deacetylation of histones (*yellow arrows*), and Ubp10 dependent deubiquitylation of H2B (*red arrows*) (and hence, prevention of H3K4me and H3K79me; see text for details), which depends on the Sir complex for proper telomere proximal targeting. On the other hand, "active chromatin" marks such as acetylation (*ac*), ubiquitylation (*ub*) and certain methylations (*me*) are sustained by cognate enzymes or enzyme complexes at the euchromatin-like regions and the boundaries. Only relevant factors depicted, not to scale

8.5 Conclusions and Perspectives

Recent studies of covalent post-translational modifications of histones not only discovered a plethora of novel modification targets (Freitas et al. 2004), but also intriguing cross-talks among these modifications, which are instrumental for the functional outcomes. In this regard, we have summarized above findings from our laboratory and others on H3S10ph in activated gene transcription and ubiquitylation/deubiquitylation of histone H2B in gene expression and silencing.

How does H3S10ph contribute to activation of gene transcription? For the cases in which H3S10ph is required for optimal H3K14ac (such as the *INO1* promoter in budding yeast), it can be argued that the downstream effects of phosphorylation on transcriptional initiation are mediated by a pathway involving SAGA (Cheung et al. 2000; Clements et al. 2003; Lo et al. 2000, 2001). On the other hand, at the promoters that are regulated by H3S10ph but without apparent link to histone H3 acetylation (e.g., heat shock genes in *Drosophila* [Labrador and Corces 2003] and *GAL1* in budding yeast [Lo et al. 2005]), the downstream effectors of H3S10ph are not known. Identification of H3S10ph as a requirement for proper TBP binding to its cognate DNA sites at promoters with both coupled and uncoupled phosphorylation/acetylation (i.e., *INO1* and *GAL1*, respectively) argues for a common downstream effector (Lo et al. 2005). This finding is also in line with the observation that the H3S10ph requirement for transcriptional activation of promoters correlates with the promoter's requirement for TBP (Labrador and Corces 2003). Further studies are required to address the universality and directness of this requirement.

We have shown that both histone H2B ubiquitylation and its deubiquitylation are important for the transcriptional activation of certain genes and maintenance of silent chromatin in budding yeast. This reversible process likely exerts its effect through trans tail cross talk by affecting histone methylation levels and in turn binding certain effector proteins in some cases, such as the Sir proteins. In this regard, SAGA transcriptional coregulator possesses a functional histone deubiquitylation (HDUB) module (i.e., Ubp8/Sgf11), in addition to a histone acetylation (HAT) module (Gcn5/Ada2/Ada3), and these have quite distinct functions within the complex (Ingvarsdottir et al. 2005; Lee et al. 2005).

Both enzymatic activities are required for SAGA's role in transcriptional activation of certain genes. One novel aspect to dynamic histone ubiquitylation is that it appears to increase early in gene activity and then, in contrast to histone acetylation, is rapidly cleaved (Henry et al. 2003; Kao et al. 2004). The failure to remove ubiquitin blocks the sequence of steps that occurs during gene activation and results in the failure to progress from H3K4me to H3K36me (Henry et al. 2003). Ubp8 is targeted to genes in a local fashion, through zinc finger domain and Sgf11-mediated association with SAGA (Ingvarsdottir et al. 2005), which is recruited to promoters through activator interaction with its Tra1 subunit (Brown et al. 2001). Another deubiquitylating enzyme, Ubp10, is involved in keeping histone H2B ubiquitylation levels (and consequently, histone methylation levels) low at the silent chromatin, leading to proper Sir protein localization and silencing (Emre et al. 2005). The two HDUBs, Ubp10 and Ubp8 are specific for their roles in silencing and activated gene expression, respectively, demonstrating a division of labor among deubiquitylating activities (Emre et al. 2005). In addition to this distinction, recent findings hint at yet additional processes where Ubp10 and Ubp8 have overlapping histone H2B deubiquitylating functions, possibly in the regulation of nontargeted global histone H2B deubiquitylation (Gardner et al. 2005). The exact identities of these additional histone targets and the physiological consequences of their (de)ubiquitylation remain to be identified.

Acknowledgements. We thank W.S. Lo, K. Ingvarsdottir, A. Wyce, K.W. Henry, A. Wood, N.J. Krogan, K. Li, R. Marmorstein, J.F. Greenblatt, A. Shilatifard, and L. Pillus for their contributions to studies reviewed here. These studies are supported by grants from NIH (GM 55360 to S.L.B) and the NSF (MCB-0078940 to S.L.B).

References

Aguilar RC, Wendland B (2003) Ubiquitin: not just for proteasomes anymore. Curr Opin Cell Biol 15:184–190

Allfrey VG, Faulkner R, Mirsky AE (1964) Acetylation and methylation of histones and their possible role in the regulation of RNA synthesis. Proc Natl Acad Sci U S A 51:786–794

Anest V, Hanson JL, Cogswell PC, Steinbrecher KA, Strahl BD, Baldwin AS (2003) A nucleosomal function for IkappaB kinase-alpha in NF-kappaB-dependent gene expression. Nature 423:659–663

Bateman A, Coin L, Durbin R, Finn RD, Hollich V, Griffiths-Jones S, Khanna A, Marshall M, Moxon S, Sonnhammer EL, Studholme DJ, Yeats C, Eddy SR (2004) The Pfam protein families database. Nucleic Acids Res 32 Database issue:D138–D141

Baur JA, Zou Y, Shay JW, Wright WE (2001) Telomere position effect in human cells. Science 292:2075–2077

Berger SL (2002) Histone modifications in transcriptional regulation. Curr Opin Genet Dev 12:142–148

Bernstein BE, Humphrey EL, Erlich RL, Schneider R, Bouman P, Liu JS, Kouzarides T, Schreiber SL (2002) Methylation of histone H3 Lys 4 in coding regions of active genes. Proc Natl Acad Sci U S A 99:8695–8700

Bettiga M, Calzari L, Orlandi I, Alberghina L, Vai M (2004) Involvement of the budding yeast metacaspase Yca1 in ubp10Delta-programmed cell death. FEMS Budding yeast Res 5:141–147

Bhaumik SR, Green MR (2001) SAGA is an essential in vivo target of the budding yeast acidic activator Gal4p. Genes Dev 15:1935–1945

Botstein D, Fink GR (1988) Budding yeast: an experimental organism for modern biology. Science 240:1439–1443

Bottomley MJ (2004) Structures of protein domains that create or recognize histone modifications. EMBO Rep 5:464–469

Boyer LA, Latek RR, Peterson CL (2004) The SANT domain: a unique histone-tail-binding module? Nat Rev Mol Cell Biol 5:158–163

Briggs SD, Bryk M, Strahl BD, Cheung WL, Davie JK, Dent SY, Winston F, Allis CD (2001) Histone H3 lysine 4 methylation is mediated by Set1 and required for cell growth and rDNA silencing in Saccharomyces cerevisiae. Genes Dev 15:3286–3295

Briggs SD, Xiao T, Sun ZW, Caldwell JA, Shabanowitz J, Hunt DF, Allis CD, Strahl BD (2002) Gene silencing: trans-histone regulatory pathway in chromatin. Nature 418:498

Brown CE, Howe L, Sousa K, Alley SC, Carrozza MJ, Tan S, Workman JL (2001) Recruitment of HAT complexes by direct activator interactions with the ATM-related Tra1 subunit. Science 292:2333–2337

Brownell JE, Zhou J, Ranalli T, Kobayashi R, Edmondson DG, Roth SY, Allis CD (1996) Tetrahymena histone acetyltransferase A: a homolog to budding yeast Gcn5p linking histone acetylation to gene activation. Cell 84:843–851

Candau R, Zhou J, Allis CD, Berger SL (1997) Histone acetyltransferase activity and interaction with ADA2 are critical for GCN5 function in vivo. EMBO J. 16:555–565

Cheung P, Tanner KG, Cheung WL, Sassone-Corsi P, Denu JM, Allis CD (2000) Synergistic coupling of histone H3 phosphorylation and acetylation in response to epidermal growth factor stimulation. Mol Cell 5:905–915

Clements A, Poux AN, Lo WS, Pillus L, Berger SL, Marmorstein R (2003) Structural basis for histone and phosphohistone binding by the GCN5 histone acetyltransferase. Mol Cell 12:461–473

Cosgrove MS, Boeke JD, Wolberger C (2004) Regulated nucleosome mobility and the histone code. Nat Struct Mol Biol 11:1037–1043

Daniel JA, Torok MS, Sun ZW, Schieltz D, Allis CD, Yates JR 3rd, Grant PA (2004) Deubiquitination of histone H2B by a budding yeast acetyltransferase complex regulates transcription. J Biol Chem 279:1867–1871

De la Cruz X, Lois S, Sanchez-Molina S, Martinez-Balbas MA (2005) Do protein motifs read the histone code? Bioessays 27:164–175

Dover J, Schneider J, Tawiah-Boateng MA, Wood A, Dean K, Johnston M, Shilatifard A (2002) Methylation of histone H3 by COMPASS requires ubiquitination of histone H2B by Rad6. J Biol Chem 277:28368–28371

Elgin SC, Workman JL (2000) Chromatin structure and gene expression, vol 35, 2nd edn. Oxford University Press, Oxford

Emre NCT, Ingvarsdottir K, Wyce A, Wood A, Krogan NJ, Henry KW, Li K, Marmorstein R, Greenblatt JF, Shilatifard A, Berger SL (2005) Maintenance of low histone ubiquitylation by Ubp10 correlates with telomere-proximal Sir2 association and gene silencing. Mol Cell 17:585–594

Fischle W, Wang Y, Allis CD (2003) Histone and chromatin cross-talk. Curr Opin Cell Biol 15:172–183

Freitas MA, Sklenar AR, Parthun MR (2004) Application of mass spectrometry to the identification and quantification of histone post-translational modifications. J Cell Biochem 92:691–700

Gardner RG, Nelson ZW, Gottschling DE (2005) Ubp10/Dot4p regulates the persistence of ubiquitinated histone H2B: distinct roles in telomeric silencing and general chromatin. Mol Cell Biol 25:6123–6139

Gasser SM, Cockell MM (2001) The molecular biology of the SIR proteins. Gene 279:1–16

Gottschling DE, Aparicio OM, Billington BL, Zakian VA (1990) Position effect at *S. cerevisiae* telomeres: reversible repression of Pol II transcription. Cell 63:751–762

Grant PA, Duggan L, Cote J, Roberts SM, Brownell JE, Candau R, Ohba R, Owen-Hughes T, Allis CD, Winston F, Berger SL, Workman JL (1997) Budding yeast Gcn5 functions in two multisubunit complexes to acetylate nucleosomal histones: characterization of an Ada complex and the SAGA (Spt/Ada) complex. Genes Dev 11:1640–1650

Grant PA, Sterner DE, Duggan LJ, Workman JL, Berger SL (1998) The SAGA unfolds: convergence of transcription regulators in chromatin-modifying complexes. Trends Cell Biol 8:193–197

Grunstein M (1997) Molecular model for telomeric heterochromatin in budding yeast. Curr Opin Cell Biol 9:383–387

Hassig CA, Fleischer TC, Billin AN, Schreiber SL, Ayer DE (1997) Histone deacetylase activity is required for full transcriptional repression by mSin3A. Cell 89:341–347

Henry KW, Wyce A, Lo WS, Duggan LJ, Emre NC, Kao CF, Pillus L, Shilatifard A, Osley MA, Berger SL (2003) Transcriptional activation via sequential histone H2B ubiquitylation and deubiquitylation, mediated by SAGA-associated Ubp8. Genes Dev 17:2648–2663

Hicke L (2001) Protein regulation by monoubiquitin. Nat Rev Mol Cell Biol 2:195–201

Hochstrasser M (1996) Ubiquitin-dependent protein degradation. Annu Rev Genet 30:405–439

Hook SS, Orian A, Cowley SM, Eisenman RN (2002) Histone deacetylase 6 binds polyubiquitin through its zinc finger (PAZ domain) and copurifies with deubiquitinating enzymes. Proc Natl Acad Sci U S A 99:13425–13430

Huang Y (2002) Transcriptional silencing in Saccharomyces cerevisiae and Schizosaccharomyces pombe. Nucleic Acids Res 30:1465–1482

Hwang WW, Venkatasubrahmanyam S, Ianculescu AG, Tong A, Boone C, Madhani HD (2003) A conserved RING finger protein required for histone H2B monoubiquitination and cell size control. Mol Cell 11:261–266

Ingvarsdottir K, Krogan NJ, Emre NC, Wyce A, Thompson NJ, Emili A, Hughes TR, Greenblatt J, Berger SL (2005) H2B ubiquitin protease Ubp8 and Sgf11 constitute a discrete functional module within the Saccharomyces cerevisiae SAGA complex. Mol Cell Biol 25:1162–1172

Kahana A, Gottschling DE (1999) DOT4 links silencing and cell growth in *Saccharomyces cerevisiae*. Mol Cell Biol 19:6608–6620

Kao CF, Hillyer C, Tsukuda T, Henry K, Berger S, Osley MA (2004) Rad6 plays a role in transcriptional activation through ubiquitylation of histone H2B. Genes Dev 18:184–195

Khorasanizadeh S (2004) The nucleosome: from genomic organization to genomic regulation. Cell 116:259–272

Kimura A, Umehara T, Horikoshi M (2002) Chromosomal gradient of histone acetylation established by Sas2p and Sir2p functions as a shield against gene silencing. Nat Genet 32:370–377

Krogan NJ, Kim M, Tong A, Golshani A, Cagney G, Canadien V, Richards DP, Beattie BK, Emili A, Boone C, Shilatifard A, Buratowski S, Greenblatt J (2003) Methylation of histone H3 by Set2 in Saccharomyces cerevisiae is linked to transcriptional elongation by RNA polymerase II. Mol Cell Biol 23:4207–4218

Kubicek S, Jenuwein T (2004) A crack in histone lysine methylation. Cell 119:903–906

Kuo MH, Allis CD (1998) Roles of histone acetyltransferases and deacetylases in gene regulation. Bioessays 20:615–626

Kurdistani SK, Grunstein M (2003) Histone acetylation and deacetylation in budding yeast. Nat Rev Mol Cell Biol 4:276–284

Labrador M, Corces VG (2003) Phosphorylation of histone H3 during transcriptional activation depends on promoter structure. Genes Dev 17:43–48

Ladurner AG, Inouye C, Jain R, Tjian R (2003) Bromodomains mediate an acetyl-histone encoded antisilencing function at heterochromatin boundaries. Mol Cell 11:365–376

Larschan E, Winston F (2001) The S. cerevisiae SAGA complex functions in vivo as a coactivator for transcriptional activation by Gal4. Genes Dev 15:1946–1956

Lee DY, Teyssier C, Strahl BD, Stallcup MR (2004) Role of protein methylation in regulation of transcription. Endocr Rev 26:147–170

Lee KK, Florens L, Swanson SK, Washburn MP, Workman JL (2005) The deubiquitylation activity of Ubp8 is dependent upon Sgf11 and its association with the SAGA complex. Mol Cell Biol 25:1173–1182

Lee KK, Prochasson P, Florens L, Swanson SK, Washburn MP, Workman JL (2004) Proteomic analysis of chromatin-modifying complexes in Saccharomyces cerevisiae identifies novel subunits. Biochem Soc Trans 32:899–903

Lo WS, Trievel RC, Rojas JR, Duggan L, Hsu JY, Allis CD, Marmorstein R, Berger SL (2000) Phosphorylation of serine 10 in histone H3 is functionally linked in vitro and in vivo to Gcn5-mediated acetylation at lysine 14. Mol Cell 5:917–926

Lo WS, Duggan L, Emre NCT, Belotserkovskya R, Lane WS, Shiekhattar R, Berger SL (2001) Snf1 – a histone kinase that works in concert with the histone acetyltransferase Gcn5 to regulate transcription. Science 293:1142–1146

Lo WS, Gamache ER, Henry KW, Yang D, Pillus L, Berger SL (2005) Histone H3 phosphorylation can promote TBP recruitment through distinct promoter-specific mechanisms. EMBO J 24:997–1008

Luger K (2003) Structure and dynamic behavior of nucleosomes. Curr Opin Genet Dev 13:127–135

Mahadevan LC, Willis AC, Barratt MJ (1991) Rapid histone H3 phosphorylation in response to growth factors, phorbol esters, okadaic acid, and protein synthesis inhibitors. Cell 65:775–783

Marcus GA, Silverman N, Berger SL, Horiuchi J, Guarente L (1994) Functional similarity and physical association between GCN5 and ADA2: putative transcriptional adaptors. EMBO J 13:4807–4815

Meneghini MD, Wu M, Madhani HD (2003) Conserved histone variant H2A.Z protects euchromatin from the ectopic spread of silent heterochromatin. Cell 112:725–736

Ng HH, Xu RM, Zhang Y, Struhl K (2002) Ubiquitination of histone H2B by Rad6 is required for efficient Dot1-mediated methylation of histone H3 lysine 79. J Biol Chem 277:34655–34657

Ng HH, Ciccone DN, Morshead KB, Oettinger MA, Struhl K (2003a) Lysine-79 of histone H3 is hypomethylated at silenced loci in budding yeast and mammalian cells: a potential mechanism for position-effect variegation. Proc Natl Acad Sci U S A 100:1820–1825

Ng HH, Robert F, Young RA, Struhl K (2003b) Targeted recruitment of set1 histone methylase by elongating Pol II provides a localized mark and memory of recent transcriptional activity. Mol Cell 11:709–719

Nishioka K, Chuikov S, Sarma K, Erdjument-Bromage H, Allis CD, Tempst P, Reinberg D (2002) Set9, a novel histone H3 methyltransferase that facilitates transcription by precluding histone tail modifications required for heterochromatin formation. Genes Dev 16:479–489

Nowak SJ, Corces VG (2004) Phosphorylation of histone H3: a balancing act between chromosome condensation and transcriptional activation. Trends Genet 20:214–220

Orlandi I, Bettiga M, Alberghina L, Vai M (2004) Transcriptional profiling of ubp10 null mutant reveals altered subtelomeric gene expression and insurgence of oxidative stress response. J Biol Chem 279:6414–6425

Osley MA (2004) H2B ubiquitylation: the end is in sight. Biochim Biophys Acta 1677:74–78

Perrod S, Gasser SM (2003) Long-range silencing and position effects at telomeres and centromeres: parallels and differences. Cell Mol Life Sci 60:2303–2318

Polioudaki H, Markaki Y, Kourmouli N, Dialynas G, Theodoropoulos PA, Singh PB, Georgatos SD (2004) Mitotic phosphorylation of histone H3 at threonine 3. FEBS Lett 560:39–44

Powell DW, Weaver CM, Jennings JL, McAfee KJ, He Y, Weil PA, Link AJ (2004) Cluster analysis of mass spectrometry data reveals a novel component of SAGA. Mol Cell Biol 24:7249–7259

Ren Q, Gorovsky MA (2001) Histone H2A.Z acetylation modulates an essential charge patch. Mol Cell 7:1329–1335

Rice JC, Briggs SD, Ueberheide B, Barber CM, Shabanowitz J, Hunt DF, Shinkai Y, Allis CD (2003) Histone methyltransferases direct different degrees of methylation to define distinct chromatin domains. Mol Cell 12:1591–1598

Robzyk K, Recht J, Osley MA (2000) Rad6-dependent ubiquitination of histone H2B in budding yeast. Science 287:501–504

Rusche LN, Kirchmaier AL, Rine J (2003) The establishment, inheritance, and function of silenced chromatin in Saccharomyces cerevisiae. Annu Rev Biochem 72:481–516

Sanders SL, Jennings J, Canutescu A, Link AJ, Weil PA (2002) Proteomics of the eukaryotic transcription machinery: identification of proteins associated with components of budding yeast TFIID by multidimensional mass spectrometry. Mol Cell Biol 22:4723–4738

Santos-Rosa H, Schneider R, Bannister AJ, Sherriff J, Bernstein BE, Emre NCT, Schreiber SL, Mellor J, Kouzarides T (2002) Active genes are tri-methylated at K4 of histone H3 Nature 419:407–411

Santos-Rosa H, Schneider R, Bernstein BE, Karabetsou N, Morillon A, Weise C, Schreiber SL, Mellor J, Kouzarides T (2003) Methylation of histone H3 K4 mediates association of the Isw1p ATPase with chromatin. Mol Cell 12:1325–1332

Santos-Rosa H, Bannister AJ, Dehe PM, Geli V, Kouzarides T (2004) Methylation of H3 lysine 4 at euchromatin promotes Sir3p association with heterochromatin. J Biol Chem 279:47506–47512

Seigneurin-Berny D, Verdel A, Curtet S, Lemercier C, Garin J, Rousseaux S, Khochbin S (2001) Identification of components of the murine histone deacetylase 6 complex: link between acetylation and ubiquitination signaling pathways. Mol Cell Biol 21:8035–8044

Shi Y, Lan F, Matson C, Mulligan P, Whetstine JR, Cole PA, Casero RA, Shi Y (2004) Histone demethylation mediated by the nuclear amine oxidase homolog LSD1. Cell 119:941–953

Sims RJ 3rd, Nishioka K, Reinberg D (2003) Histone lysine methylation: a signature for chromatin function. Trends Genet 19:629–639

Singer MS, Kahana A, Wolf AJ, Meisinger LL, Peterson SE, Goggin C, Mahowald M, Gottschling DE (1998) Identification of high-copy disruptors of telomeric silencing in Saccharomyces cerevisiae. Genetics 150:613–632

Soboleva TA, Baker RT (2004) Deubiquitinating enzymes: their functions and substrate specificity. Curr Protein Pept Sci 5:191–200

Sterner DE, Berger SL (2000) Acetylation of histones and transcription-related factors. Microbiol Mol Biol Rev 64:435–459

Struhl K (1998) Histone acetylation and transcriptional regulatory mechanisms. Genes Dev 12:599–606

Suka N, Luo K, Grunstein M (2002) Sir2p and Sas2p opposingly regulate acetylation of budding yeast histone H4 lysine16 and spreading of heterochromatin. Nat Genet 32:378–383

Sun ZW, Allis CD (2002) Ubiquitination of histone H2B regulates H3 methylation and gene silencing in budding yeast. Nature 418:104–108

Thomson S, Mahadevan LC, Clayton AL (1999) MAP kinase-mediated signalling to nucleosomes and immediate-early gene induction. Semin Cell Dev Biol 10:205–214

Turner BM (2002) Cellular memory and the histone code. Cell 111:285–291

Van Leeuwen F, Gottschling DE (2002) Genome-wide histone modifications: gaining specificity by preventing promiscuity. Curr Opin Cell Biol 14:756–762

Vaquero A, Loyola A, Reinberg D (2003) The constantly changing face of chromatin. Sci Aging Knowledge Environ 2003:RE4

Wilkinson KD (1997) Regulation of ubiquitin-dependent processes by deubiquitinating enzymes. FASEB J 11:1245–1256

Wilkinson KD (2000) Ubiquitination and deubiquitination: targeting of proteins for degradation by the proteasome. Semin Cell Dev Biol 11:141–148

Wolffe A (1998) Chromatin: structure and function, 3rd edn. Academic Press, San Diego

Wood A, Krogan NJ, Dover J, Schneider J, Heidt J, Boateng MA, Dean K, Golshani A, Zhang Y, Greenblatt JF, Johnston M, Shilatifard A (2003) Bre1, an E3 ubiquitin ligase required for recruitment and substrate selection of Rad6 at a promoter. Mol Cell 11:267–274

Wyce A, Henry KW, Berger SL (2004) H2B ubiquitylation and de-ubiquitylation in gene activation. Novartis Found Symp 259:63–73; discussion 73–77, 163–169

Xiao T, Hall H, Kizer KO, Shibata Y, Hall MC, Borchers CH, Strahl BD (2003) Phosphorylation of RNA polymerase II CTD regulates H3 methylation in budding yeast. Genes Dev 17:654–663

Yamamoto Y, Verma UN, Prajapati S, Kwak YT, Gaynor RB (2003) Histone H3 phosphorylation by IKK-alpha is critical for cytokine-induced gene expression. Nature 423:655–659

9 Histone Acetylation-Mediated Chromatin Compaction During Mouse Spermatogenesis

J. Govin, C. Lestrat, C. Caron, C. Pivot-Pajot, S. Rousseaux, S. Khochbin

Abstract. One of the most dramatic chromatin remodelling events takes place during mammalian spermatogenesis involving massive incorporation of somatic and testis-specific histone variants, as well as generalized histone modifications before their replacement by new DNA packaging proteins. Our data suggest that the induced histone acetylation occurring after meiosis may direct the first steps of genome compaction. Indeed, a double bromodomain-containing protein expressed in postmeiotic cells, Brdt, shows the extraordinary capacity to specifically condense acetylated chromatin in vivo and in vitro. In elongating spermatids, Brdt widely co-localizes with acetylated histones before accumulating in condensed chromatin domains. These domains preferentially maintain their acetylation status until late spermatogenesis. Based on these data, we propose that Brdt mediates a general histone acetylation-induced chromatin compaction and also maintains differential acetylation of specific regions, and is therefore involved in organizing the spermatozoon's genome.

9.1 Introduction

Prokaryotes and eukaryotes are capable of activating specific programs, sporulation or gametogenesis, mediating an extreme compaction of their genome. In eukaryotes, the genome compaction mechanism is preceded by meiosis generating haploid cells, which then undergo a DNA condensation process. In the animal kingdom, spermatogenesis designates the generation of male reproductive haploid cells containing a tightly packed genome in many species (Lewis et al. 2003). The compaction of the genome is stabilized through the action of specialized proteins, which replace histones transiently or permanently. Despite the fundamental importance of this event, very little is known on the molecular basis of histone replacement and genome compaction. Two hints from the literature indicate that the transitions in DNA-packaging structures are preceded by a direct alteration of chromatin structure. First, in addition to several somatic-type histone variants, many testis-specific variants are massively incorporated into chromatin. Second, a widespread histone hyperacetylation is known to occur in many vertebrate species before their replacement (Govin et al. 2004). This acetylation, which occurs in a replication- and transcription-independent manner, seems to be closely linked to histone replacement. Indeed, histones remain underacetylated in species where they persist all through spermiogenesis such as winter flounder and carp (Kennedy and Davies 1980, 1981). However, the role of the acetylation of core histones in their replacement or in other events has remained largely unknown. Some in vitro experiments suggest that histone acetylation could facilitate their displacement by protamines (Oliva et al. 1987; Oliva and Mezquita 1986), but there is no hint in the literature on how it could affect in vivo chromatin remodelling in spermatids.

The discovery of bromodomain as an acetyl-lysine binding module unveiled some of the molecular links between histone acetylation and downstream events (Yang 2004). Bromodomain defines a restricted conserved domain in a group of cellular proteins mostly involved in chromatin-related functions. Histone acetylation would therefore generate a signal recruiting a variety of functions, such as specific enzymatic activity, to chromatin (Yang 2004).

It is therefore reasonable to expect the participation of bromodomain-containing proteins also in triggering new functions after global histone acetylation in elongating spermatids. Following this reasoning, we searched databanks for testis-specific bromodomain-containing proteins susceptible of reorganizing the genome in maturing spermatids. A double bromodomain-containing protein of unknown function, Brdt, attracted our attention and was found to harbour all the properties designing it as a potentially important factor acting on acetylated chromatin in elongating spermatids (Pivot-Pajot et al. 2003). Our data indeed showed that Brdt specifically and efficiently binds acetylated histone H4 tail, and that this binding needed the integrity of both its bromodomains. The gene is specifically expressed in testis and the encoding mRNA is present in elongating spermatids where the protein could act on acetylated chromatin.

Here, we succeeded in generating antibodies specifically recognizing Brdt and monitored its expression and intracellular localization during mouse spermatogenesis. Considering its expression pattern, the unique biochemical properties of the protein, as well as its intranuclear localization in elongating spermatids, we propose that the induced histone hyperacetylation in spermatids is a preliminary signal for chromatin condensation further mediated by Brdt and analogous factors. Moreover, Brdt and related proteins may also maintain specific chromatin regions acetylated until late spermatogenesis, therefore organizing the spermatozoon's genome.

9.2 Results

9.2.1 Induced Histone Acetylation in Postmeiotic Spermatogenic Cells

Postmeiotic maturation of male germ cells has been shown in several vertebrate species to involve an enhanced acetylation of histones (Govin et al. 2004). We previously showed that this phenomenon also occurs during mouse and human spermatogenesis (Faure et al. 2003; Hazzouri et al. 2000). Here we more precisely characterized histone acetylation occurring during mouse spermatogenesis by taking advantage of a spermatogenic cell fractionation procedure. Dissociated spermatogenic cells were fractionated on a BSA gradient and fractions enriched in pachytene spermatocytes (three fractions), round (three fractions), round-elongating (one fraction) and elongating-condensing (one fraction) spermatids were obtained and histone acetylation analysed by Western blot. This analysis shows that among the core histones, mainly histone H4 and H2A are affected by the postmeiotic wave of histone hyperacetylation (Fig. 1A and data not shown).

In order to evaluate how this specific histone acetylation is associated with genome organization in maturing spermatids, we conducted a series of in situ analyses of histone H4 acetylation. Round spermatids contained essentially hypoacetylated histones (Fig. 1B). A sudden and widespread histone H4 acetylation occurred at the beginning of the elongation process, globally affecting the entire genome. As elongation proceeded the acetylation signal disappeared in a region-specific manner: a region containing compact heterochromatin-type structures, as can be judged by DAPI staining, seems to resist deacetylation/histone replacement events in elongating spermatids.

These data convey potentially important information. The induction of acetylation occurs on a very large scale without any regional specificity, therefore suggesting a sudden disequilibrium between cellular histone acetyltransferse (HAT) and histone deacetylase (HDAC) activities, and corresponding to the initiation of a chromatin compaction phase. In the subsequent phases of chromatin compaction, acetylated histone H4 becomes preferentially associated with specific regions of the spermatid genome. It is therefore extremely important to identify

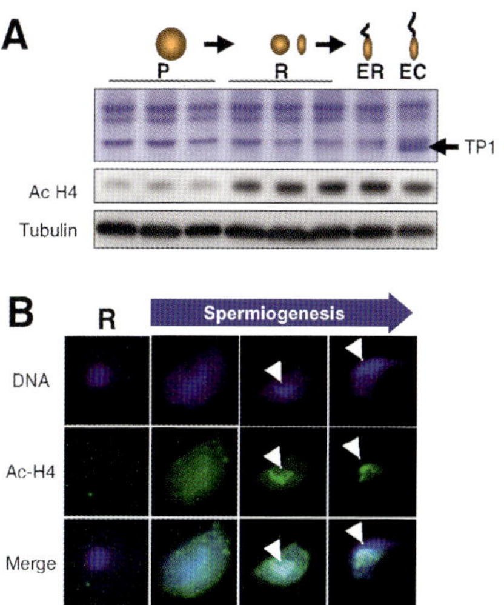

Fig. 1A,B. Induced histone H4 acetylation during spermiogenesis. **A** Mouse spermatogenic cells were fractionated on a BSA gradient and proteins in different collected fractions were analysed either by SDS-PAGE (*upper panel*, Coomassie-stained gel) or by Western blots. The fractions are enriched in meiotic spermatocytes (*P*, three fractions collected), round spermatids (*R*, three fractions collected), round and elongating spermatids (*ER*) and elongating and condensing spermatids (*EC*). The blot containing the indicated fractions was successively probed with antibodies detecting the indicated proteins. **B** Organization of acetylated H4-containing genomic regions during spermiogenesis was visualized after immunodetection of acetylated H4 on a suspension of spermatogenic cells. The *arrows* indicate compact chromatin regions containing acetylated histone H4

factors capable of interpreting the signals generated by these specific acetylations.

9.2.2 Brdt, a Unique Factor Capable of Specifically Inducing the Compaction of Acetylated Chromatin

In search of factors capable of mediating the acetylation-mediated chromatin reorganization in elongating spermatids, we previously screened available sequences in the databanks for bromodomain-containing factors specifically expressed in testis. This analysis attracted our attention to Brdt, a double bromodomain-containing protein of unknown function found expressed in the human testis (Jones et al. 1997). Our analysis showed that Brdt-encoding mRNA is also over-expressed in the mouse testis and is present in spermatocytes, as well as in spermatids where histone hyperacetylation takes place. We showed that the protein was capable of interacting with acetylated histone H4 tails and that the two bromodomains were necessary for an efficient interaction. Moreover, the protein showed the extraordinary capacity of inducing a condensation of acetylated chromatin in vivo and in vitro (Pivot-Pajot et al. 2003).

In order to further test the relationship between the Brdt-induced chromatin compaction and the accumulation of acetylated histone H4 after a tricostatin A TSA treatment, we performed an in vivo follow-up

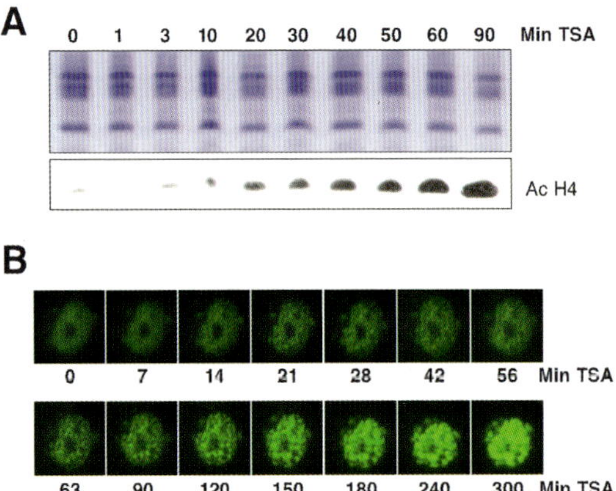

Fig. 2A,B. Time course of Brdt-dependent chromatin compaction after TSA treatment. **A** Cos cells were treated for the indicated times (min) with 100 ng/ml of TSA and the accumulation of acetylated histone H4 was monitored as in Fig. 1. The *upper panel* shows a Coomassie-stained gel of the proteins analysed in the *lower panel*. **B** Cos cells were transfected with GFP-Brdt and at the time 0 an image of a transfected cell was taken. Then 100 ng/ml of TSA was added to the culture medium. At the indicated time points after TSA addition, the GFP fluorescence of the same cell was recorded

of Brdt-induced compaction. Figure 2B shows GFP-Brdt in a single cell analysed at different times after the addition of TSA to the culture medium. After 20–30 min of TSA treatment, the first signs of chromatin compaction were clearly detectable. These time lapses also correspond to the first visible accumulation of acetylated H4 (Fig. 2A). These data confirm our previous observations and establish a strict correlation between histone acetylation and Brdt-mediated chromatin compaction.

More experiments further unravelled new aspects of Brdt- and acetylation-dependent chromatin condensation.

Histone H1 participates in the formation and stabilization of the 30-nm chromatin fibre but at the same time prevented the formation of tightly packed chromatin structures such as mitotic chromosomes (Khochbin 2001). In this regard, mitotic H1 phosphorylation is thought to loosen H1-DNA interaction, thereby allowing chromatin condensation during mitosis (Roth and Allis 1992). We wanted to know how Brdt manages to induce the formation of highly compact chromatin domains despite the presence of H1. Cos cells were transfected with GFP-Brdt, treated with TSA to induce histone hyperacetylation and endogenous histone H1° was detected using an appropriate antibody. Figure 3 shows that chromatin compaction was observed only in Brdt-expressing cells (DNA panel, compare Brdt expressing cells with nontransfected cells), as expected. Interestingly, H1° labelling was fainter in Brdt-expressing cells compared to non-transfected cells. Moreover, careful examination of the nuclei from Brdt expressing cells shows that H1° is released from chromatin as the protein fills all the intranuclear spaces devoid of DNA generated by chromatin compaction (Fig. 3, lower panel, *arrowheads*). This phenomenon was Brdt-dependent since in non-transfected cells, H1° perfectly co-localizes with chromatin (Fig. 3 and data not shown). These data show that Brdt is also capable of inducing a release of linker histones from chromatin. The removal of H1 could be facilitated by histone acetylation (Misteli et al. 2000; Perry and Annunziato 1989) as well as by Brdt-induced constraints on the chromatin fibre.

This property of Brdt may have an important implication in chromatin remodelling during spermatid maturation, namely in the exchange of linker histones. Indeed, testis specific linker histones replace somatic type H1s. H1t accumulates in meiotic spermatocytes and a specific testis-specific linker histone, Hils, becomes associated with chromatin

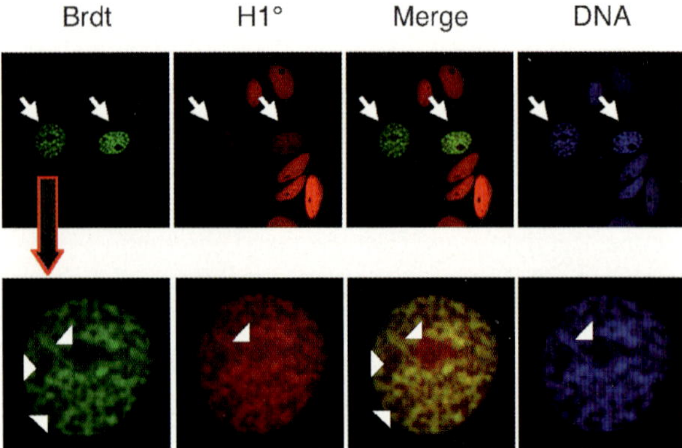

Fig. 3. Release of histone H1° from chromatin after Brdt-induced chromatin compaction. Cos cells were transfected with GFP-Brdt and then treated with 100 ng/ml of TSA for 12 h. Endogenous histone H1° was visualized in non-transfected and Brdt-expressing cells (*arrows*) after cells fixation and immunostaining. GFP fluorescence (*green*) and immunofluorescence (*red*) were recorded. The nucleus of a Brdt-expressing cell is shown in higher magnification to visualize Brdt-H1° co-localizations. *Arrowheads* show intranuclear spaces devoid of chromatin

precisely during the chromatin compaction phase in spermatids (Govin et al. 2004). Brdt, in addition to chromatin compaction, may also facilitate linker histone removal and replacement.

9.2.3 Brdt Accumulation During Mouse Spermatogenesis

Our data show that Brdt binds acetylated histones and has the capacity of inducing the compaction of acetylated chromatin. We have shown here that acetylation widely affects histone H4 at the beginning of spermatid elongation, as the whole genome seems to contain acetylated histones. However, during later stages, histone acetylation was mostly observed on dense and compact chromatin regions. We produced an antibody specifically recognizing Brdt and used it to localize the protein in spermatogenic cells. The antibody did not significantly detect Brdt in round spermatids. However, interestingly, at the

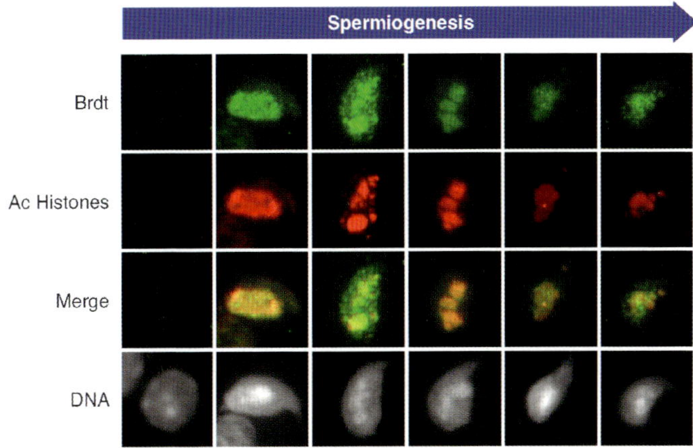

Fig. 4. Endogenous Brdt co-localizes with acetylated histones in elongating spermatids. Spermatogenic cells were recovered on a slide directly from a frozen piece of mouse testis, fixed and stained for Brdt and acetylated histones. For the purpose of co-detection, a mouse monoclonal antibody raised against acetylated lysines (Komatsu et al. 2003) was used to detect acetylated histones. Brdt was detected using an immunopurified polyclonal antibody raised in our laboratory

beginning of elongation, an intense Brdt signal appears, which perfectly co-localizes with acetylated histones (Fig. 4). During later stages, acetylated histones become preferentially visible on condensed DNA regions and Brdt continues to perfectly co-localize with these regions of compact chromatin. These observations are in agreement with a model attributing to Brdt a role in the condensation of acetylated chromatin domains.

9.3 Discussion

Basic testis-specific proteins have been thought for a long time to be responsible for the genome compaction during the maturation of spermatids or spermiogenesis. In mammals these basic proteins are transition proteins (TPs; Meistrich et al. 2003) and protamines (Lewis et al. 2003). TPs arrive first and are then replaced by protamines. Interestingly, the generation of mice bearing invalidated TP1/2 genes showed that the first steps of chromatin condensation can take place in the total absence of TPs (Shirley et al. 2004; Zhao et al. 2004) indicating that another yet unidentified mechanism initiates chromatin compaction. Our data suggest for the first time that the massive histone acetylation may unexpectedly mediate this chromatin compaction. Indeed, the timing of induced histone acetylation during mouse spermiogenesis corresponds to the initiation spermatid elongation and genome condensation (Hazzouri et al. 2000). Moreover, our analysis showed that Brdt is also expressed in elongating spermatids. Brdt has the unique property of inducing the condensation of acetylated chromatin. In fact, we were able to mimic, in somatic cells, the situation observed in spermatids, i.e. Brdt expression in the presence of hyperacetylated histones. Histone acetylation was induced after the inhibition of histone deacetylases (HDACs) in Brdt-expressing cells in culture. Remarkably, Brdt mediated a dramatic chromatin compaction strictly dependent on histone acetylation. This experiment showed that when both Brdt and acetylated histones are present at the same time in a cell, the protein is capable of inducing a chromatin compaction. These conditions are precisely found in elongating spermatids, suggesting therefore that histone acetylation and Brdt would mediate the first chromatin condensation observed during spermiogenesis before the synthesis of TPs. This hypothesis is also supported by the fact that in situ analysis showed that Brdt perfectly co-localizes with acetylated chromatin in the spermatids at various stages of genome compaction. Based on these data, we could reasonably propose that chromatin acetylation is a preliminary signal for genome compaction and that Brdt is one of the actors interpreting the signal.

Our investigations also revealed a new property of Brdt. Indeed, Brdt-mediated compaction of chromatin was associated with the release of linker histones. It is therefore possible that Brdt also mediates the exchange of linker histones taking place in spermatids as somatic type H1s are replaced by H1t and Hils.

During spermatogenesis, Brdt probably ensures multiple functions. Its accumulation at the time of histone hyperacetylation would mediate the first steps of chromatin compaction. The protein could, at the same time, favour the exchange of linker histones. Finally, Brdt could also protect specific chromatin region containing acetylated histones against their replacement by TPs. Along this line, it has been shown that the yeast homologue of Brdt, Bdf1, is capable of protecting acetylated H4 against deacetylation and actively maintains acetylated chromatin regions (Ladurner et al. 2003). Likewise, Brdt could also actively maintain specific chromatin region in acetylated state. In agreement with this hypothesis, our in situ analysis showed that Brdt and acetylated histones co-localized in specific domains during late spermatogenesis when most of the histones are replaced in other regions of the genome. This observation suggests that Brdt may be involved in establishing the functional organization of the spermatozoon's genome. In mammals, spermatozoa contain variable amounts of histones. The human sperm nucleus retains about 15% of the usual haploid histone complement and also contains a heterogeneous mixture of protamines (Balhorn et al. 1987) and other basic proteins (Ammer et al. 1986; Balhorn et al. 1987; Gusse et al. 1986; Tanphaichitr et al. 1978). Electron microscopy of spread human sperm chromatin has resolved globular structures, which correspond in size to nucleosomes (Gusse and Chevaillier 1980; Sobhon et al. 1982a, 1982b; Wagner and Yun 1981). Gatewood et al. (Gatewood et al. 1987) have demonstrated that the nucleohistone and nucleoprotamine portions of human sperm chromatin form discrete structures, which are, at least partly, sequence-specific (Wykes and Krawetz 2003). Based on these observations, it is tempting to propose that the acetylation-dependent recruitment of Brdt may also prevent the assembly of TPs and protamines on specific regions and therefore participate in organizing the spermatozoon's genome. It would also be very interesting to investigate the function of these particular regions during early embryogenesis. One prediction would be that these regions would transmit some male-specific information to the embryo.

Brdt is a member of BET double bromodomain-containing proteins also sharing a remarkably well-conserved domain of unknown function named ET (extra terminal). The family encompasses three somatic members known as Brd2, Brd3 and Brd4 in addition to the testis-specific member Brdt (Florence and Faller 2001; Shang et al. 2004). Several studies have established that different members may present different acetylated histone-binding activities. Brd2 specifically interacts with histone H4 acetylated on K12 and also shows a weak binding to H2B acetylated on K5 and K12 (Kanno et al. 2004). Brd4 interacts with acetylated H4 and H3 tails while, under the same conditions, Brd2 does not interact with acetylated histone H3 (Dey et al. 2003). Both Brd4 and Brd2 have the unique properties of remaining bound to chromatin of condensed mitotic chromosomes when other bromodomain-containing proteins were released (Dey et al. 2003). One important function of these proteins could therefore be the transmission of the acetylation mark to daughter cells. Indeed, a massive histone deacetylation takes place during mitosis, which corresponds to the erasure of the histone acetylation-dependent epigenetic information (Kruhlak et al. 2001). The binding of Brd2 and Brd4 to acetylated nucleosomes would therefore first protect specific chromatin regions against deacetylation during mitosis and then ensure the transmission of this acetylation mark to the daughter cells. The ability of Brd4 to bind mitotic chromosomes has also been used by the bovine papillomavirus to tether its genome to the host cell mitotic chromosomes and to therefore ensure the transmission of the viral episomes to the daughter cells (You et al. 2004).

There is now strong evidence on the participation of these double Brd proteins in tumorigenesis, mainly in the development of B-cell lymphoma and leukaemia. Human leukaemic cell lines or primary leukaemic blasts have high levels of Brd2, and the transgenic mice model expressing Brd2 in lymphoid cells develops aggressive B-cell lymphoma showing all the characteristics of human non-Hodgkin lymphoma (Greenwald et al. 2004).

In a highly lethal form of epithelial tumour, the poorly differentiated carcinoma, a chromosome translocation t(15;19)(q13, p13.1) causes the fusion of Brd4 to the novel testis-specific protein NUT (nuclear protein in testis), resulting in a fusion oncogene and deregulation of the activities of both proteins (French et al. 2003).

Finally, the aberrant expression of Brdt has been reported in 12 of 47 cases of non-small cell lung cancers (Scanlan et al. 2000). Considering the unique properties of Brdt, it is expected that its illegitimate expression in somatic cells would have deleterious effects. Brdt would locally induce the compaction of acetylated chromatin and this phenomenon may have a dramatic effect on gene expression and its illegitimate expression may represent an important risk factor for cancer development.

9.4 Material and Methods

9.4.1 Spermatogenic Cell Fractionation

A cell fractionation procedure was performed as previously described (Pivot-Pajot et al. 2003). Briefly, after euthanasia, the testes of two adult mice were dissected and the albuginea removed. The seminiferous tubules were released by incubation in a collagenase solution. The germinal and Sertoli cells were then released from the tubules by pipetting. The cells were finally resuspended in a volume of 18 ml of the sedimentation solution and laid on a 360-ml 2%–4% BSA gradient in an airtight sedimentation chamber at 4 °C. The cells were allowed to sediment at 4 °C for 70 min. Ten-millilitre fractions were then collected on ice and centrifuged. The cells present in each collected fraction were identified by observation under a phase contrast microscope, and the fractions were pooled in order to obtain cell suspensions enriched in pachytene spermatocytes (P), round and elongating spermatids (RES) or condensing spermatids (CS). Each cell suspension was then washed again in phosphate buffered saline at 4 °C and processed for further analysis.

9.4.2 Cell Lines, Transfection and Cell Treatment

Cos7 cells were cultured in Dulbecco's modified Eagle medium (Gibco, Invitrogen, San Diego, CA, USA) supplemented with penicillin, streptomycin and 10% FCS (Gibco). Cells at 50% confluence in one-well coverslips were transfected with 2 μg of δC-sBrdt (Pivot-Pajot et al. 2003) using the FuGENE 6 reagent (Roche, Meylan, France). Twelve hours after transfection, cells were treated with Trichostatin A (100 ng/ml) and used for further analysis.

9.4.3 Immunofluorescence

Somatic Cell Preparations

Transfected cells (Cos7) grown on coverslips were fixed in 4% paratormaldehyde for 1 min at room temperature, and then permeabilized in 4% paraformaldehyde, 0.1% Triton-X100 for 1 min at room temperature.

Testis Imprint Preparations

Mouse testes were cut into three and frozen in liquid nitrogen. Testis imprint were prepared by gently pressing the frozen testis tissue onto glass slides, air drying them, incubating them in 90% ethanol for 3 min, and air drying them again. They were then permeabilized in Saponine 0.5%/Triton 0.25%/phosphate buffered saline for 15 min at room temperature.

Immunofluorescent Staining

Following incubation with the blocking buffer (10% skimmed milk and 0.2% Tween in phosphate buffered saline) for 30 min, primary antibody diluted in 1% skimmed milk and 0.2% Tween in phosphate buffered saline was added on the slides for 1 h at 37 °C. Primary antibodies were a monoclonal antibody against H1° (Gorka et al. 1998), a rabbit polyclonal antibody against acetylated H4 diluted at 1:100 (Upstate #06-598), a monoclonal antibody against acetylated lysine diluted at 1:250 (Komatsu et al. 2003), and an immunopurified rabbit anti-BRDT polyclonal antibody. The slides were then washed three times in 1% skimmed milk and 0.2% Tween in phosphate buffered saline and further incubated for 30 min at 37 °C with anti-mouse IgG and/or anti-rabbit IgG (Molecular Probe, Invitrogen, San Diego, CA, USA) conjugated with Alexa 488 or 456, respectively. After three washes in 1% skimmed milk and 0.2% Tween in phosphate buffered saline, the cells were counterstained with Hoechst (250 ng/ml) and examined under a fluorescence microscope.

9.4.4 Microscopy

Cell preparations were first analysed with an epifluorescence microscope (Axiophot, Zeiss) using a 63× oil immersion objective. For 3D analysis, a confocal laser scanning microscope (LSM 510-NLO, Zeiss), equipped with a 40× oil immersion objective (NA = 1.3), was used. Following excitation at the appropriate wavelengths (at 488 nm using an argon ion laser, at 543 nm with a HeNe laser, and at 750 nm with a Ti:Sa laser (SpectraPhysics, Mountain View, CA, USA) of the Alexa 488, 546 and Hoechst signals, stacks of equidistant 0.1 μm of 128 × 128 pixel size images (with a zoom of 3.9×, one pixel = 0.1 μm) were recorded. Images were averaged eight times.

9.4.5 In Vivo Cell Imaging

Cos7 cells growing on coverslips were transfected with δC-sBrdt-GFP 24 h before imaging. Cells were then transferred into an incubation chamber (37 °C, 5% CO_2) of a videomicroscopy station. TSA was added to the cell culture medium at the final concentration of 100 ng/ml, and images were acquired every 7 min using a Zeiss Axiovert 100 M microscope with an air 40× objective controlled by Metamorph software.

9.4.6 Western Blots

Samples were submitted to electrophoresis on a 15% SDS-polyacrylamide gel and transferred to nitrocellulose membranes (Hybond C+, Amersham Biosciences, Buckinghamshire, UK) according to standard procedures. Following a 30 min incubation in phosphate-buffered saline containing 10% skimmed milk and 0.3% Tween (blocking solution), the blots were incubated for 1 h at room temperature with primary antibody diluted in phosphate-buffered saline, containing 10% FCS and 0.2% Tween. After three washes of 5 min each at room temperature, the blots were incubated with horseradish peroxidase-conjugated secondary antibody in phosphate-buffered saline containing 1% dry milk. The final detection was performed using the ECL+ kit (Amersham).

Acknowledgements. We are also grateful to Sandrine Curtet-Benitski for technical assistance. This work was supported by the Ligue National Contre le Cancer, Comité de l'Isère and the Région Rhône-Alpes Emergence program. CL is recipient of a Region Rhône-Alpes PhD fellowship. SR is recipient of an INSERM interface program.

References

Ammer H, Henschen A, Lee CH (1986) Isolation and amino-acid sequence analysis of human sperm protamines P1 and P2. Occurrence of two forms of protamine P2. Biol Chem Hoppe Seyler 367:515–522

Balhorn R, Corzett M, Mazrimas J, Stanker LH, Wyrobek A (1987) High-performance liquid chromatographic separation and partial characterization of human protamines 1, 2, and 3. Biotechnol Appl Biochem 9:82–88

Dey A, Chitsaz F, Abbasi A, Misteli T, Ozato K (2003) The double bromodomain protein Brd4 binds to acetylated chromatin during interphase and mitosis. Proc Natl Acad Sci U S A 100:8758–8763

Faure AK, Pivot-Pajot C, Kerjean A, Hazzouri M, Pelletier R, Peoc'h M, Sele B, Khochbin S, Rousseaux S (2003) Misregulation of histone acetylation in Sertoli cell-only syndrome and testicular cancer. Mol Hum Reprod 9:757–763

Florence B, Faller DV (2001) You bet-cha: a novel family of transcriptional regulators. Front Biosci 6:D1008–D1018

French CA, Miyoshi I, Kubonishi I, Grier HE, Perez-Atayde AR, Fletcher JA (2003) BRD4-NUT fusion oncogene: a novel mechanism in aggressive carcinoma. Cancer Res 63:304–307

Gatewood JM, Cook GR, Balhorn R, Bradbury EM, Schmid CW (1987) Sequence-specific packaging of DNA in human sperm chromatin. Science 236:962–964

Gorka C, Brocard MP, Curtet S, Khochbin S (1998) Differential recognition of histone H10 by monoclonal antibodies during cell differentiation and the arrest of cell proliferation. J Biol Chem 273:1208–1215

Govin J, Caron C, Lestrat C, Rousseaux S, Khochbin S (2004) The role of histones in chromatin remodelling during mammalian spermiogenesis. Eur J Biochem 271:3459–3469

Greenwald RJ, Tumang JR, Sinha A, Currier N, Cardiff RD, Rothstein TL, Faller DV, Denis GV (2004) E mu-BRD2 transgenic mice develop B-cell lymphoma and leukemia. Blood 103:1475–1484

Gusse M, Chevaillier P (1980) Electron microscope evidence for the presence of globular structures in different sperm chromatins. J Cell Biol 87:280–284

Gusse M, Sautiere P, Belaiche D, Martinage A, Roux C, Dadoune JP, Chevaillier P (1986) Purification and characterization of nuclear basic proteins of human sperm. Biochim Biophys Acta 884:124–134

Hazzouri M, Pivot-Pajot C, Faure AK, Usson Y, Pelletier R, Sele B, Khochbin S, Rousseaux S (2000) Regulated hyperacetylation of core histones during mouse spermatogenesis: involvement of histone deacetylases. Eur J Cell Biol 79:950–960

Jones MH, Numata M, Shimane M (1997) Identification and characterization of BRDT: a testis-specific gene related to the bromodomain genes RING3 and Drosophila fsh. Genomics 45:529–534

Kanno T, Kanno Y, Siegel RM, Jang MK, Lenardo MJ, Ozato K (2004) Selective recognition of acetylated histones by bromodomain proteins visualized in living cells. Mol Cell 13:33–43

Kennedy BP, Davies PL (1980) Acid-soluble nuclear proteins of the testis during spermatogenesis in the winter flounder. Loss of the high mobility group proteins. J Biol Chem 255:2533–2539

Kennedy BP, Davies PL (1981) Phosphorylation of a group of high molecular weight basic nuclear proteins during spermatogenesis in the winter flounder. J Biol Chem 256:9254–9259

Khochbin S (2001) Histone H1 diversity: bridging regulatory signals to linker histone function. Gene 271:1–12

Komatsu Y, Yukutake Y, Yoshida M (2003) Four different clones of mouse anti-acetyllysine monoclonal antibodies having different recognition properties share a common immunoglobulin framework structure. J Immunol Methods 272:161–175

Kruhlak MJ, Hendzel MJ, Fischle W, Bertos NR, Hameed S, Yang XJ, Verdin E, Bazett-Jones DP (2001) Regulation of global acetylation in mitosis through loss of histone acetyltransferases and deacetylases from chromatin. J Biol Chem 276:38307–38319

Ladurner AG, Inouye C, Jain R, Tjian R (2003) Bromodomains mediate an acetyl-histone encoded antisilencing function at heterochromatin boundaries. Mol Cell 11:365–376

Lewis JD, Abbott DW, Ausio J (2003) A haploid affair: core histone transitions during spermatogenesis. Biochem Cell Biol 81:131–140

Meistrich ML, Mohapatra B, Shirley CR, Zhao M (2003) Roles of transition nuclear proteins in spermiogenesis. Chromosoma 111:483–488

Misteli T, Gunjan A, Hock R, Bustin M, Brown DT (2000) Dynamic binding of histone H1 to chromatin in living cells. Nature 408:877–881

Oliva R, Bazett-Jones D, Mezquita C, Dixon GH (1987) Factors affecting nucleosome disassembly by protamines in vitro. Histone hyperacetylation and chromatin structure, time dependence, and the size of the sperm nuclear proteins. J Biol Chem 262:17016–17025

Oliva R, Mezquita C (1986) Marked differences in the ability of distinct protamines to disassemble nucleosomal core particles in vitro. Biochemistry 25:6508–6511

Perry CA, Annunziato AT (1989) Influence of histone acetylation on the solubility, H1 content and DNase I sensitivity of newly assembled chromatin. Nucleic Acids Res 17:4275–4291

Pivot-Pajot C, Caron C, Govin J, Vion A, Rousseaux S, Khochbin S (2003) Acetylation-dependent chromatin reorganization by BRDT, a testis-specific bromodomain-containing protein. Mol Cell Biol 23:5354–5365

Roth SY, Allis CD (1992) Chromatin condensation: does histone H1 dephosphorylation play a role? Trends Biochem Sci 17:93–98

Scanlan MJ, Altorki NK, Gure AO, Williamson B, Jungbluth A, Chen YT, Old LJ (2000) Expression of cancer-testis antigens in lung cancer: definition of bromodomain testis-specific gene (BRDT) as a new CT gene, CT9. Cancer Lett 150:155–164

Shang E, Salazar G, Crowley TE, Wang X, Lopez RA, Wolgemuth DJ (2004) Identification of unique, differentiation stage-specific patterns of expression of the bromodomain-containing genes Brd2, Brd3, Brd4, Brdt in the mouse testis. Gene Expr Patterns 4:513–519

Shirley CR, Hayashi S, Mounsey S, Yanagimachi R, Meistrich ML (2004) Abnormalities and reduced reproductive potential of sperm from Tnp1- and Tnp2-null double mutant mice. Biol Reprod 71:1220–1229

Sobhon P, Chutatape C, Chalermisarachai P, Vongpayabal P, Tanphaichitr N (1982a) Transmission and scanning electron microscopic studies of the human sperm chromatin decondensed by micrococcal nuclease and salt. J Exp Zool 221:61–79

Sobhon P, Tanphaichitr N, Chutatape C, Vongpayabal P, Panuwatsuk W (1982b) Electron microscopic and biochemical analyses of the organization of human sperm chromatin decondensed with sarkosyl and dithiothreitol. J Exp Zool 223:277–290

Tanphaichitr N, Sobhon P, Taluppeth N, Chalermisarachai P (1978) Basic nuclear proteins in testicular cells and ejaculated spermatozoa in man. Exp Cell Res 117:347–356

Wagner TE, Yun JS (1981) Human sperm chromatin has a nucleosomal structure. Arch Androl 7:251–257

Wykes SM, Krawetz SA (2003) The structural organization of sperm chromatin. J Biol Chem 278:29471–29477

Yang XJ (2004) Lysine acetylation and the bromodomain: a new partnership for signaling. Bioessays 26:1076–1087

You J, Croyle JL, Nishimura A, Ozato K, Howley PM (2004) Interaction of the bovine papillomavirus E2 protein with Brd4 tethers the viral DNA to host mitotic chromosomes. Cell 117:349–360

Zhao M, Shirley CR, Mounsey S, Meistrich ML (2004) Nucleoprotein transitions during spermiogenesis in mice with transition nuclear protein Tnp1 and Tnp2 mutations. Biol Reprod 71:1016–1025

10 Role of Ubiquitin-Like Proteins in Transcriptional Regulation

R.T. Hay

Abstract. Conjugation of ubiquitin-like proteins (Ubls) to components of the transcriptional machinery represents an important mechanism to allow switching between different activity states. While ubiquitin modification of transcription factors is associated with transcriptional activation, SUMO modification of transcription factors is most often associated with transcriptional repression. Recent experiments indicate that another Ubl, NEDD8, can also influence transcription. One of the characteristics of Ubl modification is that the biological consequences of conjugation do not appear proportionate to the small fraction of substrate that is modified. The low steady state levels of Ubl-modified substrates can be attributed to a highly dynamic situation in which proteins are conjugated to a particular Ubl only for the modification to be removed by Ubl-specific proteases. It therefore appears that an unmodified protein with a history of Ubl modification may have different properties from a protein that never has been modified. Here the diverse effects of Ubl modification are discussed and models proposed to explain Ubl actions.

10.1 The Family of Small Ubiquitin-Like Modifiers

Ubiquitin and its relatives the ubiquitin-like proteins (Ubls) are ligated to proteins, altering the properties of the modified proteins and hugely increasing the complexity of the proteome in eukaryotic cells. One of the differences between modification with small chemical groups such as phosphate or acetate and the Ubls is that modification with the Ubl effectively creates a new protein with additional interfaces for protein–protein interactions. It is likely that the new interfaces created by Ubls on modified proteins are responsible for their diverse biological effects. Like modification with small chemical groups, modifications with Ubls are reversible with a conserved set of enzymes being responsible for Ubl addition and a family of proteases being responsible for their removal. In addition to ubiquitin, *s*mall *u*biquitin-like *mo*difier (SUMO) and *n*eural precursor cell *e*xpressed *d*evelopmentally *d*own-regulated gene 8 (NEDD8) (Kamitani et al. 1997) or *r*elated to *ub*iquitin 1 (Rub 1) in yeast can be covalently coupled to target proteins with important functional consequences (Jentsch and Pyrowolakis 2000). Ubls are conjugated to target proteins by enzymatic cascades that typically involve three activities (Fig. 1): an E1 (activating enzyme), E2 (conjugating enzyme), and E3 (protein ligase) (Huang et al. 2004). Although NEDD8 is highly homologous to ubiquitin (57% identity), it has unique E1 and E2 enzymes that discriminate between ubiquitin and NEDD8. The NEDD8 E1 is a heterodimer composed of the amyloid precursor protein-binding protein (APP-BP1) and the Uba3 protein, while the NEDD8 E2 is Ubc12 (Gong et al. 1999; Lammer et al. 1998; Liakopoulos et al. 1998; Osaka et al. 1998; Pozo et al. 1998). The small ubiquitin-like modifier SUMO-1 is rather distantly related to ubiquitin and NEDD8, and has many biological functions, including control of gene expression, maintenance of genome integrity, intracellular transport and protein stability (reviewed in Verger et al. 2003). SUMO-1 is also known as SMT3c, PIC1, GMP1, sentrin and Ublp1, and the pathway of SUMO conjugation is required for cell viability in yeast and higher eukaryotes (Hayashi et al. 2002;

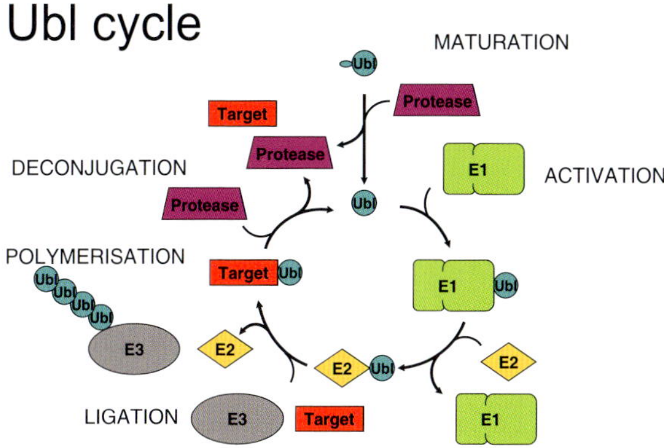

Fig. 1. Ubl modification. The Ubl precursor is processed by a Ubl-specific protease to reveal the C-terminal diglycine that is activated by the E1 enzyme. Following transesterification onto the E2 conjugating enzyme, the protein target is selected and with the help of an E3 ligase, the Ubl is ligated to the substrate. The Ubl can be deconjugated from the target protein by the action of Ubl-specific proteases

Seufert et al. 1995). In lower eukaryotes, such as yeast, insects and nematodes, only one SUMO gene is expressed (more commonly termed Smt3), whereas in plants up to eight versions of SUMO are expressed (Kurepa et al. 2003). In vertebrates, three paralogues, designated SUMO-1, SUMO-2 and SUMO-3 are expressed. SUMO-2 (also known as Smt3a and Sentrin3) and SUMO-3 (also known as Smt3b and Sentrin2), only differ from one another by three N-terminal residues and have yet to be functionally differentiated (Saitoh and Hinchey 2000; Tatham et al. 2001). They form a distinct subfamily known as SUMO-2/-3 and are 50% identical in sequence to SUMO-1.

The SUMO activating enzyme is a heterodimer containing SAE1 and SAE2 subunits (known as AOS1 and UBA2 in yeast) (Desterro et al. 1999; Gong et al. 1999; Johnson et al. 1997; Okuma et al. 1999) that both contain regions homologous to the ubiquitin activating enzyme. To initiate the SUMO modification reaction, SAE1/SAE2 utilizes ATP and SUMO and catalyses the formation of adenylated SUMO in which the C-terminal carboxyl group of SUMO is covalently linked to AMP. Pyrophosphate is released from the enzyme and breakage of the SUMO-AMP linkage results in the formation of a covalent intermediate in which the C terminal carboxyl group of SUMO forms a thioester bond with the sulphydryl group of a cysteine residues in SAE2 (C173). In the second step of the reaction, SUMO is transesterified from SAE2 to cysteine 93 in the SUMO conjugating enzyme Ubc9 (Desterro et al. 1997; Johnson and Blobel 1997). One of the features of Ubc9 that distinguishes it from conjugating enzymes of other ubiquitin-like proteins is its ability to directly recognize substrate proteins. Thus, Ubc9-SUMO thioester can catalyse formation of an isopeptide bond between the C-terminal carboxyl group of SUMO and the ε-amino group of lysine in the substrate protein, provided that the lysine is part of a SUMO conjugation motif. Typically, lysine residues subject to SUMO modification are found within a SUMO modification consensus motif (Fig. 1), ψKxE (where ψ is a large hydrophobic residue and x any residue) (Rodriguez et al. 2001) that is directly contacted by Ubc9 (Bernier-Villamor et al. 2002; Lin et al. 2002; Sampson et al. 2001). However, additional residues close to the active site of Ubc9 may have a role in correctly orienting the C-terminus of the thioester linked SUMO to facilitate formation of the isopeptide bond (Tatham et al. 2003). In the presence of only SAE1/SAE2 and Ubc9 SUMO is specifically conjugated to substrates containing the ψKxE motif. However, with the notable exception of RanGAP1, this reaction is rather inefficient and additional components are clearly required to accelerate this reaction (Lin et al. 2002; Tatham et al. 2003). In the ubiquitin system, substrate recognition is mediated by a family of E3 protein ligases that range in complexity from a single polypeptide species to the multicomponent anaphase promoting complex (APC). These ubiquitin E3 ligases all function by direct and specific recognition of the substrate protein and either deliver ubiquitin to substrate from an internal thioester (for example E6-AP) or act as an adapter to bring together substrate and ubiquitin thioester-loaded E2 (for example SCF complexes). The discovery of proteins which increase the efficiency of SUMO conjugation has indicated the existence of SUMO E3 ligases, and while they all appear to specifically bind Ubc9, their interaction with substrate appears much less specific. Although these E3 ligases accelerate the rate of SUMO modification in vitro and in vivo in a substrate-specific fashion, they appear to target multiple proteins with few recognizably similar features. In higher eukaryotes, proteins with SUMO E3 ligase activity are found in a number of different functional groups. The PIAS proteins constitute a family of SUMO regulators including PIAS1, PIASxα, PIASxβ, PIASγ and PIAS3 which, like the Siz proteins in yeast, contain a region homologous to the RING domains of ubiquitin E3 ligases (Hochstrasser 2001). While the Siz/PIAS family of SUMO E3 ligases are similar to RING containing ubiquitin E3s, the SUMO-specific ligases RanBP2 and Polycomb group protein 2 (Pc2) are unrelated to the ubiquitin E3s. RanBP2 is located at nuclear pore complexes and accelerates the SUMO modification of a variety of substrates (Kirsh et al. 2002; Pichler et al. 2002). RanBP2 does contain a RING-like domain, but it is without the region required for E3 activity, and the interaction of RanBP2 with Ubc9 indicates that it functions very differently from RING or HECT domain containing E3 liagases (Pichler et al. 2004; Tatham et al. 2004). Pc2 is a component of the Polycomb chromatin modifying complex whose activity results in transcriptional repression. The SUMO E3 ligase activity of Pc2 is manifest on the C-terminal binding protein (CtBP) transcriptional co-repressor, and binding of CtBP and Ubc9 by Pc2 results in their sequestration into subnuclear domains known as PcG bodies (Kagey et al. 2003).

10.2 SUMO-1 Modification and Transcription

Proteins modified by SUMO are not restricted to a particular functional class and the effects of SUMO modification vary from substrate to substrate. However, a large proportion of the reported SUMO targets are involved in the control of transcription, and these have been catalogued in a number of recent reviews (Girdwood et al. 2004; Muller et al. 2004; Seeler and Dejean 2003). SUMO modification can positively or negatively influence transcription. In a limited number of cases, SUMO modification correlates with increased transcriptional activity. SUMO modification of heat-shock transcription factors HSF1 and HSF2 increases DNA-binding activity, and blocking modification by mutation decreases HSF1 transcriptional activity (Goodson et al. 2001; Hong et al. 2001). Similarly, SUMO modification of p53 leads to an increase in transcriptional and apoptotic responses (Gostissa et al. 1999; Rodriguez et al. 1999). However, in most cases SUMO modification is associated with transcriptional repression. Thus the sites of SUMO modification have been mapped to previously identified repression domains in transcription factors, and mutation of lysine acceptor residues results in increased transcriptional activity. However, genetic analysis in *Caenorhabditis elegans* has recently demonstrated the physiological role of SUMO modification in mediating transcriptional repression of the Hox genes. Polycomb group (PcG) protein SOP-2 directly interacts with Ubc9 via its conserved SAM domain and as a consequence is modified by SUMO. SUMO modification of SOP-2 is required for its ability to repress Hox genes, and disruption of SUMO modification results in ectopic expression of Hox genes and homeotic transformations. SUMO modification of SOP-2 results in its recruitment into subnuclear bodies, and these data indicate that recruitment into these bodies is required for transcriptional repression (Zhang et al. 2004).

Androgen receptor (AR) is modified by SUMO, on a lysine acceptor within a classical SUMO conjugation motif, and mutation of the target lysine residue resulted in an increase in transcriptional output, indicating a negative role for SUMO in AR-dependent transcription (Poukka et al. 2000). The promoter regions of genes often contain multiple binding sites for single or multiple transcription factors. In such situations the bound factors have a synergistic effect on transcription. These experiments led to the identification of a novel protein domain, which appeared to limit transcriptional synergy of multiple promoter bound factors (Iniguez-Lluhi and Pearce 2000) and were identical to the previously characterized SUMO conjugation motifs. Mutations of the lysine acceptors for SUMO conjugation increased transcriptional synergy, indicating that SUMO conjugation could function to limit the extent of transcriptional synergy. Transcriptional synergy control motifs are located in other nuclear receptors such as the glucocorticoid, mineralcorticoid, and progesterone receptors, all of which are SUMO-modified. An example where modification with SUMO-1 or SUMO-2/-3 has distinct functional consequences is evident in the control of transcriptional synergy, where SUMO-2 appears to act as a more potent inhibitor at synergy control motifs in the glucocorticoid receptor than SUMO-1 (Holmstrom et al. 2003). AR is modified by SUMO at two sites and it has recently become apparent that modification by SUMO at each site has different consequences. While mutation at both sites derepresses AR-dependent transcription, only one site effects the co-operativity of the AR on multiple AREs. Remarkably, this SUMO-dependent effect on co-operativity depends on whether the response element contains inverted or direct repeats of the AR-binding site. Thus SUMO modification of AR differentially modulates co-operativity, depending on the relative orientation of the AR dimer bound to DNA (Callewaert et al. 2004). Recently a new PIAS-like protein, hZimp10, was shown to enhance the transcriptional activity of AR. hZimp10 co-localized with AR and SUMO-1 in replication foci during S phase and increased SUMO modification of AR. The ability of hZimp10 to enhance AR-dependent transcription was dependent on its ability to effect SUMO modification of AR. Thus hZimp10 is a co-activator and SUMO E3 ligase that appears to be required for full transcriptional activity of AR (Sharma et al. 2003).

p300 is a transcriptional regulator involved in many signalling pathways and co-ordinates transcription by recruiting transcriptional components to the promoter or by using its histone acetyl transferase activity to alter the histone code. Deletion analysis identified a cell cycle regulated domain (CRD1) in p300 that could repress transcription from particular promoters (Snowden et al. 2000). CRDI is located close to the bromo domain and contained two SUMO modification motifs in tandem. p300-dependent transcriptional repression was demonstrated to be dependent on SUMO modification of both motifs (Girdwood et al. 2003). Transcription factor Sp3 was also demonstrated to contain a repression domain that contained a consensus site for SUMO modification (Ross et al. 2002; Sapetschnig et al. 2002). Mutation of the SUMO acceptor lysine resulted in a dramatic derepression of transcription, although the mechanism of repression has yet to be clarified. Ross et al. (2002) reported that SUMO modification altered the subnuclear distribution of Sp3, although in the other report SUMO dependent differences in subnuclear localization were not detected (Sapetschnig et al. 2002). The transcription factor Elk-1 also contains a conserved repression domain or R-motif. The R motif is interchangeable with the CRD1 domain of p300, and it also contains a SUMO conjugation motif that is necessary for transcriptional repression (Yang et al. 2003a). The SUMO modification status of Elk-1 is regulated by the ERK MAP kinase signal transduction pathway (Yang et al. 2003b) and phosphorylation of Elk-1 by MAP kinase results in the removal of SUMO that presumably involves the recruitment of a SUMO-specific protease.

10.3 NEDD8 Modification and Transcription

The p53 tumour suppressor protein is a short-lived protein, due to its rapid proteasomal degradation. Upon exposure of cells to various stress stimuli, levels of p53 rise as a consequence of reduced proteolytic degradation and its function as a transcription factor is activated. An important regulator of the stability/activity of p53 is the Mdm2 oncogene product. Biochemical studies showed that Mdm2 can act as an E3 ubiquitin ligase, and this activity depends on its RING finger domain at the C-terminus of the protein. Mdm2 directly interacts with p53, promoting its ubiquitination and degradation through the 26S proteasome (Rodriguez et al. 2000). Although Mdm2 is recognized as a ubiquitin E3 ligase, Mdm2 also promoted conjugation of NEDD8 to p53 both in vivo and in vitro. Mdm2 is also modified with NEDD8, and this reaction displays very similar characteristics to the auto-ubiquitination activity of Mdm2. Thus, Mdm2 appears to function as an E3 ligase for the NEDDylation of p53. Mdm2-mediated NEDDylation of p53 requires three lysines (K370, K372, K373) at the C-terminus of the protein, ubiquitination requires lysines K370, K372, K373, K381, K382, K386. This difference in the lysine requirement for NEDD8 and ubiquitin conjugation of p53 indicates the specificity of Mdm2 as an E3 ligase for these two conjugation pathways. A p53 mutant unable to be modified by NEDD8 displayed an increased transcriptional activity compared to wild type p53 in H1299 cells, suggesting that NEDDylation of p53 inhibits its function as a transcription factor (Xirodimas et al. 2004).

10.4 Transcriptional Control by Ubl-Specific Proteases

In the ubiquitin system, proteins involved in processing ubiquitin and deconjugating ubiquitin from substrates are important components of the control mechanism that regulates the availability of "ready-to-conjugate" ubiquitin and the modification status of individual proteins. Likewise, ubiquitin-like processing enzymes (ULPs), are important determinants of SUMO modification status. In *S. cerevisiae*, two Smt3-specific proteases, Ulp1 and Ulp2, have been characterized and are detected at the nuclear pore and nucleoplasm, respectively (Li and Hochstrasser 1999, 2000; Takahashi et al. 2000). Both enzymes can deconjugate Smt3 from modified proteins and process Smt3 precursors to the mature form with the C-terminal diglycine. Comparison of the Ulp1 and Ulp2 sequences reveals that the region of homology is confined to a protease domain of about 200 amino acids located in the C-terminal region of the protein. SUMO-specific proteases are cysteine proteases that belong to the family of proteases typified by the adenovirus protease with a characteristic His/Asp/Cys catalytic triad. Database searching initially identified seven genes for human proteins with significant sequence homology to yeast Ulp1 that were believed to be SUMO-specific proteases (Yeh et al. 2000). Of these genes SENP1, SENP2 (also designated Axam, SuPr-1, SSP3, SMT3IP2), and SMT3IP1 have been shown to function as SUMO-specific proteases (Best et al. 2002; Girdwood et al. 2003; Hang and Dasso 2002; Kadoya et al. 2002; Nishida et al. 2000; Zhang et al. 2002). If transcription factors can be repressed by conjugation to SUMO then it follows that removal of SUMO by the action of SUMO-specific proteases would derepress transcription and have an important role in the regulation of SUMO-dependent gene expression. SUMO-dependent repression of p300, Sp3 and Elk-1 was relieved by co-expression of SUMO-specific protease SSP3/SuPr-1 (Girdwood et al. 2003; Ross et al. 2002; Yang et al. 2003a), thus suggesting a means by which these proteases could control transcription. By changing the sub-cellular localization or activity of SSP3, signalling pathways could induce deconjugation of specific SUMO-modified transcription factors and thereby activate transcription in response to specific stimuli.

While the products of the seven genes homologous to Ulp1 probably function as proteases for ubiquitin-like proteins they are not all specific for SUMO, as one member of the group has recently been revealed as a NEDD8-specific protease, NEDP1/DEN1 (Gan-Erdene et al. 2003; Mendoza et al. 2003; Wu et al. 2003). We have determined that NEDP1 is capable of deNEDDylating p53 in vivo and thus affecting the transcriptional output from p53. Recently we have determined the structure of NEDP1 alone and NEDP1 bound to NEDD8 and demonstrated that the protease undergoes a dramatic conformational change upon NEDD8 binding.

10.5 How Do Ubls Alter Transcription?

There are a number of possible models to explain Ubl-dependent transcriptional repression. In one scenario, it is proposed that Ubl-modified transcription factors recruit co-repressor molecules to susceptible promoters that bring about changes in chromatin structure associated with repression. An alternative proposal is that the Ubl-modified transcription factors are recruited into the repressive environment of particular subnuclear domains. In fact, there is evidence to support both proposals and the models are certainly not mutually exclusive. Evidence for co-repressor recruitment has come from a number of recent studies. In the case of p300, transcriptional repression is relieved by the histone deacetylase inhibitor trichostatin A, and the SUMO-modified CRD1 domain binds to HDAC6. Thus SUMO-dependent transcriptional repression is mediated by recruitment of HDAC6 to the CRD1 domain of p300 (Girdwood et al. 2003). Recruitment of HDACs to chromatin-bound p300 would result in deacetylation of core histones and other transcriptional factors, which could in turn initiate a chain of events leading to changes to chromatin structure and inhibition of transcription. HDAC-6 is a member of the class II HDACs (-4, -5, -6 and -7), a characteristic of which is the ability to shuttle between the nucleus and the cytoplasm (Khochbin et al. 2001). Therefore nuclear translocation of HDAC6 in combination with SUMO conjugation to the CRD1 domain may act as a mechanism to control the transcriptional output of p300. Histone H4 is also modified by SUMO (Shiio and Eisenman 2003) and the modification is associated with transcriptional repression. Again, this appeared to be a result of co-repressor binding with HDAC1 and HP1-γ being recruited to chromatin by SUMO modified histone H4. Recently it has been demonstrated that SUMO modification of Elk-1 results in the recruitment of histone deacetylase activity to promoters (Yang and Sharrocks 2004). In this study, chromatin immunoprecipitation assays were used to provide evidence that the SUMO-modified Elk-1 recruited HDAC-2 to responsive promoters. This is an important point as it provides direct evidence that SUMO-modified Elk-1 functions by recruiting a co-repressor to the promoter rather than SUMO modification sequestering the transcription factor away from the promoter.

An alternative mechanism to explain the repressive effect of SUMO on transcription is that SUMO-modified proteins are recruited into the repressive environment of particular subnuclear domains, the best known of which is the PML nuclear body. PML can regulate transcription by its association with numerous SUMO-modified transcription factors/cofactors such as homeodomain interacting protein kinase 2 (HIPK2), Daxx, and Sp100. SUMO modification of HIPK2 appears to be required for its recruitment into PML nuclear bodies (Kim et al. 1999), where it associates with Groucho co-repressor and HDAC1 (Choi et al. 1999). Daxx can recruit HDACs (Li et al. 2000), which may facilitate the assembly of repressive structures within the environment of the PML body (Lehembre et al. 2001; Li et al. 2000). The PML body component Sp100 contains a domain that binds heterochromatin protein 1 (HP1) (Lehming et al. 1998; Seeler et al. 1998), and the interaction with HP1α in vitro is enhanced by SUMO modification of SP100 (Seeler et al. 2001). HP1α has a well established role in initiating the formation of repressive domains in chromatin and localization of such proteins to subnuclear domains by SUMO modification of nuclear body components represents a reasonable mechanism for the creation of a locally repressive environment. Another group of proteins that have a well-established role in transcriptional repression are the Polycomb group (PcG) proteins and they have the ability to recruit proteins into the repressive environment of Polycomb nuclear bodies. Recent evidence, discussed earlier in the review, indicates that SUMO modification is required for polycomb-dependent transcriptional repression. In *C. elegans*, the PcG protein SOP-2 requires SUMO modification for its recruitment into PcG subnuclear bodies and for its ability to negatively regulate HOX gene expression (Zhang et al. 2004). In human cells, the Polycomb protein Pc2 acts as a SUMO E3 ligase, engaging both CtBP and Uc9 and mediating SUMO modification of CtBP which is required for its recruitment into Polycomb subnuclear domains and for transcriptional repression (Kagey et al. 2003).

An alternative way in which SUMO could alter transcription is by competing with other Ubls for modification of key lysine residues. This paradigm for SUMO action was revealed by studies on the NF-κB inhibitor IκBα (Desterro et al. 1998). In response to activation of a number of signal transduction pathways, IκBα undergoes polyubiquitination at lysines 21 and 22, targeting the protein for proteasome-mediated degradation. The released NFκB can then enter the nucleus, where it binds to responsive promoters in a variety of genes (Hay et al. 1999). SUMO modification of IκBα appears to stabilize the protein, as it becomes resistant to ubiquitin-mediated degradation (Desterro et al. 1998). As the lysine acceptor for SUMO-1 modification is also Lys21, SUMO and ubiquitin therefore compete for the same lysine acceptor site. Thus SUMO modification protects IκBα from signal-induced degradation, remaining bound to NF-κB and creating a "privileged pool" of IκBα that is refractile to degradation.

Competition between SUMO and ubiquitin modification is also apparent on the IκB kinase (IKK) regulator, NF-κB essential modulator (NEMO/IKKγ). In response to DNA damage signals, SUMO is conjugated to NEMO on a single lysine residue that, like IκBα, is also the site of ubiquitination. In this case, however, ubiquitination of NEMO does not target it for degradation but translocates it back to the cytoplasm after DNA damage induced, SUMO-dependent nuclear retention (Hay 2004; Huang et al. 2003).

10.6 Models for Ubl Action

While the above models may go some way toward explaining the roles of Ubls in transcriptional control, they require further elaboration to accommodate two apparently conflicting observations. In most situations, the proportion of a particular transcription factor that is Ubl-modified at the steady state is small in relation to the total pool of the transcription factor. In the case of SUMO, the transcription factor appears to be maximally repressed and mutation of the lysine residue used for SUMO modification relieves repression. To accommodate these observations, we have developed a general model that is depicted in Fig. 2. In this hypothesis, newly synthesized transcription factor is constitutively conjugated to SUMO and incorporated into a repression complex in a SUMO-dependent fashion. In the presence of SUMO-specific pro-

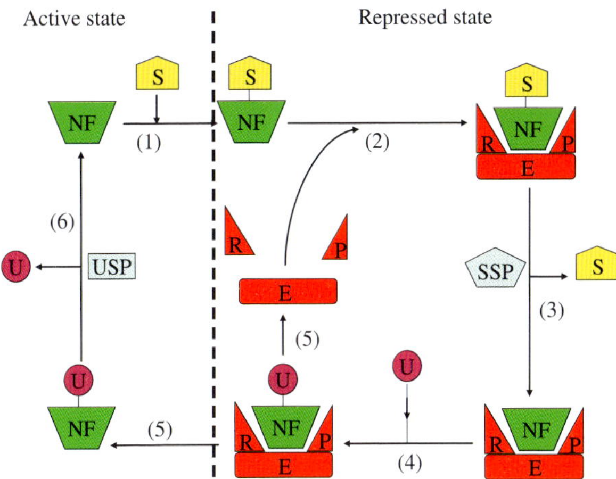

Fig. 2. Model to explain how Ubl modification acts to transfer a transcription factor from the active to the repressed state. A newly synthesized nuclear transcription factor (*NF*) is modified by SUMO (1) and is assembled (2) into a repression complex (*REP*) by the action of SUMO-specific assembly factors. Once the repression complex has been formed, the assembly factor dissociates and the SUMO can be deconjugated (3) by the action of a SUMO-specific protease (*SSP*). The transcription factor in the complex can be modified by a different ubiquitin-like protein (*U*) or another modification (4) that facilitates disruption of the complex and release of the active transcription factor (5). This may involve the action of an additional disassembly factor (not shown). Return to the unmodified state would require the deconjugation of the transcription factor (6) by enzymes such as ubiquitin-specific proteases (*USP*)

teases, the SUMO could be deconjugated, but the transcription factor would be retained in the repression complex in a SUMO-independent fashion. Slow dissociation of the repression complex could release sufficient quantities of unmodified factors to allow basal transcription. Thus SUMO is required to initiate repression, but is not required for maintenance of repression. Co-localization of enzymes involved in SUMO conjugation and deconjugation within the cell nucleus (Zhang et al. 2002) indicates that SUMO modification and deconjugation may be part of a highly dynamic process where substrates undergo rapid modification by SUMO followed by equally rapid deconjugation.

One way in which SUMO-modified proteins could be incorporated into repression complexes is that SUMO-modified proteins may interact with chaperonins that assemble the modified protein into a stable complex. Once assembled, the chaperonins can dissociate from the complex and SUMO can be deconjugated, leaving the once modified protein locked into the repressed state. Examples where specific proteins are involved in assembly but not maintenance of multiprotein complexes are well documented. During DNA replication, helicase loaders incorporate monomeric helicase units into ring-shaped hexameric helicases that encircle the DNA. The loaders utilize ATP to assemble the helicase around DNA, but after ATP hydrolysis they dissociate from the complex, leaving the helicase locked on to the DNA (Davey and O'Donnell 2003). Although SUMO modification has an important role in transcription, it has many other functions, and the hypothesis could be extended to explain other actions of SUMO. Thus SUMO conjugation may simply recruit factors required for assembly or disassembly of various macromolecular complexes. Once the modified protein has been incorporated into, or released from, the complex SUMO could be removed. This model is equally applicable to the other ubiquitin-like proteins and could simply be envisaged as a means of switching proteins between different states. Once the switch has taken place, the ubiquitin-like protein can be deconjugated and the once modified protein would remain locked into that particular state.

Once the transcriptionally repressed state has been established, it is probable that mechanisms will exist for reactivation. Transcriptional activation could be triggered by signal-induced disassembly of the complex (for example by phosphorylation or acetylation), thus releasing the active transcription factor. This could be coupled with an increased rate of SUMO deconjugation that would halt formation of new repression complexes. This seems to be the situation with Elk-1, where signal-induced phosphorylation of Elk-1 via the MAP kinase pathway results in a concomitant loss of SUMO modification on Elk-1 and activation of transcription (Yang et al. 2003a). Elk-1 phosphorylation could induce structural changes in Elk-1 that allow escape from the "repression complex", and may allow recruitment of SUMO-specific proteases to deconjugate ElK-1 and block the constitutive recruitment of Elk-1 into new repression complexes.

Lysine residues can act as acceptors for many different modifications (for example, SUMO, NEDD8 ubiquitin, methylation and acetylation), and it is therefore possible that particular lysine residues could act as convergence points for a number of different modification pathways (Freiman and Tjian 2003) to allow assembly of the modified proteins into higher-order structures with unique properties. One of the implications of this type of model is that otherwise identical unmodified protein molecules may have different properties depending on whether or not they have a history of modification. Thus a protein that has never been SUMO modified and not assembled into a SUMO-dependent repression complex could be transcriptionally active, while an identical protein that has undergone SUMO modification and incorporated into a repression complex, but has since been deconjugated could be transcriptionally inactive.

Acknowledgements. Work in the author's laboratory is supported by MRC, BBSRC, CRUK and Wellcome Trust.

References

Bernier-Villamor V, Sampson DA, Matunis MJ, Lima CD (2002) Structural basis for E2-mediated SUMO conjugation revealed by a complex between ubiquitin-conjugating enzyme Ubc9 and RanGAP1. Cell 108:345–356

Best JL, Ganiatsas S, Agarwal S, Changou A, Salomoni P, Shirihai O, Meluh PB, Pandolfi PP, Zon LI (2002) SUMO-1 protease-1 regulates gene transcription through PML Mol Cell 10:843–855

Callewaert L, Verrijdt G, Haelens A, Claessens F (2004) Differential effect of small ubiquitin-like modifier (SUMO)-ylation of the androgen receptor in the control of cooperativity on selective versus canonical response elements. Mol Endocrinol 18:1438–1449

Choi CY, Kim YH, Kwon HJ, Kim Y (1999) The homeodomain protein NK-3 recruits Groucho and a histone deacetylase complex to repress transcription. J Biol Chem 274:33194–33197

Davey MJ, O'Donnell M (2003) Replicative helicase loaders: ring breakers and ring makers. Curr Biol 13:R594–R596

Desterro JM, Thomson J, Hay RT (1997) Ubch9 conjugates SUMO but not ubiquitin. FEBS Lett 417:297–300

Desterro JM, Rodriguez MS, Hay RT (1998) SUMO-1 modification of Ikappa-Balpha inhibits NF-kappaB activation. Mol Cell 2:233–239

Desterro JM, Rodriguez MS, Kemp GD, Hay RT (1999) Identification of the enzyme required for activation of the small ubiquitin-like protein SUMO-1. J Biol Chem 274:10618–10624

Freiman RN, Tjian R (2003) Regulating the regulators: lysine modifications make their mark. Cell 112:11–17

Gan-Erdene T, Nagamalleswari K, Yin L, Wu K, Pan ZQ, Wilkinson KD (2003) Identification and characterization of DEN1, a deneddylase of the ULP family. J Biol Chem 278:28892–28900

Girdwood D, Bumpass D, Vaughan OA, Thain A, Anderson LA, Snowden AW, Garcia-Wilson E, Perkins ND, Hay RT (2003) p300 transcriptional repression is mediated by SUMO modification. Molecular Cell 18:1043–1054

Girdwood DW, Tatham MH, Hay RT (2004) SUMO and transcriptional regulation. Semin Cell Dev Biol 15:201–210

Gong L, Li B, Millas S, Yeh ET (1999) Molecular cloning and characterization of human AOS1 and UBA2, components of the sentrin-activating enzyme complex. FEBS Lett 448:185–189

Goodson ML, Hong Y, Rogers R, Matunis MJ, Park-Sarge OK, Sarge KD (2001) Sumo-1 modification regulates the DNA binding activity of heat shock transcription factor 2, a promyelocytic leukemia nuclear body associated transcription factor. J Biol Chem 276:18513–18518

Gostissa M, Hengstermann A, Fogal V, Sandy P, Schwarz SE, Scheffner M, Del Sal G (1999) Activation of p53 by conjugation to the ubiquitin-like protein SUMO-1. EMBO J 18:6462–6471

Hang J, Dasso M (2002) Association of the Human SUMO-1 Protease SENP2 with the nuclear pore. J Biol Chem 277:19961–19966

Hay RT (2004) Modifying NEMO. Nat Cell Biol 6:89–91

Hay RT, Vuillard L, Desterro JM, Rodriguez MS (1999) Control of NF-kappa B transcriptional activation by signal induced proteolysis of I kappa B alpha. Philos Trans R Soc Lond B Biol Sci 354:1601–1609

Hayashi T, Seki M, Maeda D, Wang W, Kawabe Y, Seki T, Saitoh H, Fukagawa T, Yagi H, Enomoto T (2002) Ubc9 is essential for viability of higher eukaryotic cells. Exp Cell Res 280:212–221

Hochstrasser M (2001) SP-RING for SUMO: new functions bloom for a ubiquitin-like protein. Cell 107:5–8

Holmstrom S, Van Antwerp ME, Iniguez-Lluhi JA (2003) Direct and distinguishable inhibitory roles for SUMO isoforms in the control of transcriptional synergy. Proc Natl Acad Sci U S A 100:15758–15763

Hong Y, Rogers R, Matunis MJ, Mayhew CN, Goodson M, Park-Sarge OK, Sarge KD (2001) Regulation of heat shock transcription factor 1 by stress-induced SUMO-1 modification. J Biol Chem 276:40263–40267

Huang DT, Walden H, Duda D, Schulman BA (2004) Ubiquitin-like protein activation. Oncogene 23:1958–1971

Huang TT, Wuerzberger-Davis SM, Wu ZH, Miyamoto S (2003) Sequential modification of NEMO/IKKgamma by SUMO-1 and ubiquitin mediates NF-kappaB activation by genotoxic stress. Cell 115:565–576

Iniguez-Lluhi JA, Pearce D (2000) A common motif within the negative regulatory regions of multiple factors inhibits their transcriptional synergy. Mol Cell Biol 20:6040–6050

Jentsch S, Pyrowolakis G (2000) Ubiquitin and its kin: how close are the family ties? Trends Cell Biol 10:335–342

Johnson ES, Blobel G (1997) Ubc9p is the conjugating enzyme for the ubiquitin-like protein Smt3p. J Biol Chem 272:26799–26802

Johnson ES, Schwienhorst I, Dohmen RJ, Blobel G (1997) The ubiquitin-like protein Smt3p is activated for conjugation to other proteins by an Aos1p/Uba2p heterodimer. EMBO J 16:5509–5519

Kadoya T, Yamamoto H, Suzuki T, Yukita A, Fukui A, Michiue T, Asahara T, Tanaka K, Asashima M, Kikuchi A (2002) Desumoylation activity of Axam, a novel Axin-binding protein, is involved in downregulation of beta-catenin. Mol Cell Biol 22:3803–3819

Kagey MH, Melhuish TA, Wotton D (2003) The polycomb protein Pc2 is a SUMO E3. Cell 113:127–137

Kamitani T, Kito K, Nguyen HP, Yeh ET (1997) Characterization of NEDD8, a developmentally down-regulated ubiquitin-like protein. J Biol Chem 272:28557–28562

Khochbin S, Verdel A, Lemercier C, Seigneurin-Berny D (2001) Functional significance of histone deacetylase diversity. Curr Opin Genet Dev 11:162–166

Kim YH, Choi CY, Kim Y (1999) Covalent modification of the homeodomain-interacting protein kinase 2 (HIPK2) by the ubiquitin-like protein SUMO-1. Proc Natl Acad Sci U S A 96:12350–12355

Kirsh O, Seeler JS, Pichler A, Gast A, Muller S, Miska E, Mathieu M, Harel-Bellan A, Kouzarides T, Melchior F, Dejean A (2002) The SUMO E3 ligase RanBP2 promotes modification of the HDAC4 deacetylase. EMBO J 21:2682–2691

Kurepa J, Walker JM, Smalle J, Gosink MM, Davis SJ, Durham TL, Sung DY, Vierstra RD (2003) The small ubiquitin-like modifier (SUMO) protein modification system in Arabidopsis. Accumulation of SUMO1 and -2 conjugates is increased by stress. J Biol Chem 278:6862–6872

Lammer D, Mathias N, Laplaza JM, Jiang W, Liu Y, Callis J, Goebl M, Estelle M (1998) Modification of yeast Cdc53p by the ubiquitin-related protein rub1p affects function of the SCFCdc4 complex. Genes Dev 12:914–926

Lehembre F, Muller S, Pandolfi PP, Dejean A (2001) Regulation of Pax3 transcriptional activity by SUMO-1-modified PML Oncogene 20:1–9

Lehming N, Le Saux A, Schuller J, Ptashne M (1998) Chromatin components as part of a putative transcriptional repressing complex. Proc Natl Acad Sci U S A 95:7322–7326

Li H, Leo C, Zhu J, Wu X, O'Neil J, Park EJ, Chen JD (2000) Sequestration and inhibition of Daxx-mediated transcriptional repression by PML Mol Cell Biol 20:1784–1796

Li SJ, Hochstrasser M (1999) A new protease required for cell-cycle progression in yeast. Nature 398:246–251

Li SJ, Hochstrasser M (2000) The yeast ULP2 (SMT4) gene encodes a novel protease specific for the ubiquitin-like Smt3 protein. Mol Cell Biol 20:2367–2377

Liakopoulos D, Doenges G, Matuschewski K, Jentsch S (1998) A novel protein modification pathway related to the ubiquitin system. EMBO J 17:2208–2214

Lin D, Tatham MH, Yu B, Kim S, Hay RT, Chen Y (2002) Identification of a substrate recognition site on ubc9. J Biol Chem 277:21740–21748

Mendoza HM, Shen LN, Botting C, Lewis A, Chen J, Ink B, Hay RT (2003) NEDP1, a highly conserved cysteine protease that deNEDDylates Cullins. J Biol Chem 278:25637–25643

Muller S, Ledl A, Schmidt D (2004) SUMO: a regulator of gene expression and genome integrity. Oncogene 23:1998–2008

Nishida T, Tanaka H, Yasuda H (2000) A novel mammalian Smt3-specific isopeptidase 1 (SMT3IP1) localized in the nucleolus at interphase. Eur J Biochem 267:6423–6427

Okuma T, Honda R, Ichikawa G, Tsumagari N, Yasuda H (1999) In vitro SUMO-1 modification requires two enzymatic steps E1 and E2. Biochem Biophys Res Commun 254:693–698

Osaka F, Kawasaki H, Aida N, Saeki M, Chiba T, Kawashima S, Tanaka K, Kato S (1998) A new NEDD8-ligating system for cullin-4A. Genes Dev 12:2263–2268

Pichler A, Gast A, Seeler JS, Dejean A, Melchior F (2002) The nucleoporin RanBP2 has SUMO1 E3 ligase activity. Cell 108:109–120

Pichler A, Knipscheer P, Saitoh H, Sixma TK, Melchior F (2004) The RanBP2 SUMO E3 ligase is neither HECT- nor RING-type. Nat Struct Mol Biol 11:984–991

Poukka H, Karvonen U, Janne OA, Palvimo JJ (2000) Covalent modification of the androgen receptor by small ubiquitin-like modifier 1 (SUMO-1). Proc Natl Acad Sci U S A 97:14145–14150

Pozo JC, Timpte C, Tan S, Callis J, Estelle M (1998) The ubiquitin-related protein RUB1 and auxin response in Arabidopsis. Science 280:1760–1763

Rodriguez MS, Desterro JM, Lain S, Midgley CA, Lane DP, Hay RT (1999) SUMO-1 modification activates the transcriptional response of p53. EMBO J 18:6455–6461

Rodriguez MS, Desterro JM, Lain S, Lane DP, Hay RT (2000) Multiple C-terminal lysine residues target p53 for ubiquitin-proteasome-mediated degradation. Mol Cell Biol 20:8458–8467

Rodriguez MS, Dargemont C, Hay RT (2001) SUMO-1 conjugation in vivo requires both a consensus modification motif and nuclear targeting. J Biol Chem 276:12654–12659

Ross S, Best JL, Zon LI, Gill G (2002) SUMO-1 modification represses Sp3 transcriptional activation and modulates its subnuclear localization. Mol Cell 10:831–842

Saitoh H, Hinchey J (2000) Functional heterogeneity of small ubiquitin-related protein modifiers SUMO-1 versus SUMO-2/3. J Biol Chem 275:6252–6258

Sampson DA, Wang M, Matunis MJ (2001) The small ubiquitin-like modifier-1 (SUMO-1) consensus sequence mediates Ubc9 binding and is essential for SUMO-1 modification. J Biol Chem 276:21664–21669

Sapetschnig A, Rischitor G, Braun H, Doll A, Schergaut M, Melchior F, Suske G (2002) Transcription factor Sp3 is silenced through SUMO modification by PIAS1. EMBO J 21:5206–5215

Seeler JS, Dejean A (2003) Nuclear and unclear functions of SUMO Nat Rev Mol Cell Biol 4:690–699

Seeler JS, Marchio A, Sitterlin D, Transy C, Dejean A (1998) Interaction of SP100 with HP1 proteins: a link between the promyelocytic leukemia-associated nuclear bodies and the chromatin compartment. Proc Natl Acad Sci U S A 95:7316–7321

Seeler JS, Marchio A, Losson R, Desterro JM, Hay RT, Chambon P, Dejean A (2001) Common properties of nuclear body protein SP100 and TIF1alpha chromatin factor: role of SUMO modification. Mol Cell Biol 21:3314–3324

Seufert W, Futcher B, Jentsch S (1995) Role of a ubiquitin-conjugating enzyme in degradation of S- and M-phase cyclins. Nature 373:78–81

Sharma M, Li X, Wang Y, Zarnegar M, Huang CY, Palvimo JJ, Lim B, Sun Z (2003) hZimp10 is an androgen receptor co-activator and forms a complex with SUMO-1 at replication foci. EMBO J 22:6101–6114

Shiio Y, Eisenman RN (2003) Histone sumoylation is associated with transcriptional repression. Proc Natl Acad Sci U S A 100:13225–13230

Snowden AW, Anderson LA, Webster GA, Perkins ND (2000) A novel transcriptional repression domain mediates p21(WAF1/CIP1) induction of p300 transactivation. Mol Cell Biol 20:2676–2686

Takahashi Y, Mizoi J, Toh EA, Kikuchi Y (2000) Yeast Ulp1, an Smt3-specific protease, associates with nucleoporins. J Biochem (Tokyo) 128:723–725

Tatham MH, Jaffray E, Vaughan OA, Desterro JM, Botting CH, Naismith JH, Hay RT (2001) Polymeric chains of sumo-2 and sumo-3 are conjugated to protein substrates by sae1/sae2 and ubc9. J Biol Chem 276:35368–35374

Tatham MH, Chen Y, Hay RT (2003) Role of two residues proximal to the active site of Ubc9 in substrate recognition by the Ubc9.SUMO-1 thiolester complex. Biochemistry 42:3168–3179

Tatham MH, Kim S, Jaffray E, Song J, Chen Y, Hay RT (2004) Unique binding interactions among Ubc9, SUMO and RanBP2 reveal a mechanism for SUMO paralog selection. Nat Struct Mol Biol 12:67–74

Verger A, Perdomo J, Crossley M (2003) Modification with SUMO A role in transcriptional regulation. EMBO Rep 4:137–142

Wu K, Yamoah K, Dolios G, Gan-Erdene T, Tan P, Chen A, Lee CG, Wei N, Wilkinson KD, Wang R, Pan ZQ (2003) DEN1 is a dual function protease capable of processing the C-terminus of Nedd8 deconjugating hyperneddylated CUL1. J Biol Chem 278:28882–28891

Xirodimas DP, Saville MK, Bourdon JC, Hay RT, Lane DP (2004) Mdm2-mediated NEDD8 conjugation of p53 inhibits its transcriptional activity. Cell 118:83–97

Yang SH, Sharrocks AD (2004) SUMO promotes HDAC-mediated transcriptional repression. Mol Cell 13:611–617

Yang S-H, Jaffray E, Hay RT, Sharrocks AD (2003a) Dynamic interplay of the SUMO and ERK pathways in regulating Elk-1 transcriptional activity. Mol Cell 12:63–74

Yang SH, Jaffray E, Senthinathan B, Hay RT, Sharrocks AD (2003b) SUMO and Transcriptional Repression: Dynamic Interactions between the MAP kinase and SUMO pathways. Cell Cycle 2:528–530

Yeh ET, Gong L, Kamitani T (2000) Ubiquitin-like proteins: new wines in new bottles. Gene 248:1–14

Zhang H, Saitoh H, Matunis MJ (2002) Enzymes of the SUMO modification pathway localize to filaments of the nuclear pore complex. Mol Cell Biol 22:6498–6508

Zhang H, Smolen GA, Palmer R, Christoforou A, van den Heuvel S, Haber DA (2004) SUMO modification is required for in vivo Hox gene regulation by the Caenorhabditis elegans Polycomb group protein SOP-2. Nat Genet 36:507–511

11 Interplay of the SUMO and MAP Kinase Pathways

S.-H. Yang, A.D. Sharrocks

Abstract. The SUMO modification pathway has been linked with controlling the activity of numerous transcriptional regulatory proteins. In the majority of substrates studied so far, sumoylation imparts repressive properties. In several cases, part of this mechanism has been shown to be due to SUMO-dependent recruitment of histone deacetylases (HDACs). This is exemplified by the transcription factor Elk-1, where HDAC-2 is specifically recruited in response to sumoylation. Importantly, activation of the ERK MAP kinase pathway leads to Elk-1 desumoylation and HDAC loss. Furthermore, PIAS proteins can regulate the activities of transcription factors in SUMO-dependent and -independent manners. Further links between the MAP kinase pathways and PIAS proteins have been uncovered, suggesting a complex interplay been the MAP kinase and SUMO modification pathways. Here we discuss the current evidence suggesting links between the SUMO and MAP kinase pathways and point to other potential regulatory events and how these might be affected in cancer.

11.1 Introduction

The modification of proteins by SUMO (small ubiquitin-like modifier protein) is becoming increasingly recognized as an important event. In particular, recent studies have focused on an important role for SUMO in controlling nuclear events with a particular emphasis on transcription. Several different classes of proteins involved in transcription have been shown to be modified and regulated by SUMO modification, including gene-specific transcription factors, general transcription factors, coactivators and corepressors (reviewed in Gill 2003, 2004; Verger et al. 2003). Sumoylation of proteins is known to affect many different molecular activities, including protein stability, nuclear-cytoplasmic shuttling, subnuclear localization, DNA binding, and transcriptional activation potential. However, one theme that has recently emerged is that sumoylation of transcriptional regulatory proteins often imparts repressive properties. We are now beginning to understand the molecular basis to how repression is mediated by sumoylation although protein-specific mechanisms seem to exist. Furthermore, although we know that many proteins are modified by SUMO and in many cases we know the functional consequences, little is known about how SUMO modification is regulated and hence the activity of the modified transcription factors. This review summarizes our current understanding of how an important class of cellular signalling pathways, the MAP kinase cascades, impact on transcription factor regulation by the SUMO pathway.

11.2 The SUMO Pathway

Protein sumoylation is achieved through a pathway that is analogous to the ubiquitin pathway (reviewed in Hay 2001; Muller et al. 2001). The basic pathway consists of an E1 SUMO activating enzyme, SAE1/2, and an E2 SUMO conjugating enzyme, Ubc9 (Fig. 1). These enzymes are sufficient for substrate sumoylation in vitro. However, a series of E3 ligases have recently been found which enhance substrate sumoylation.

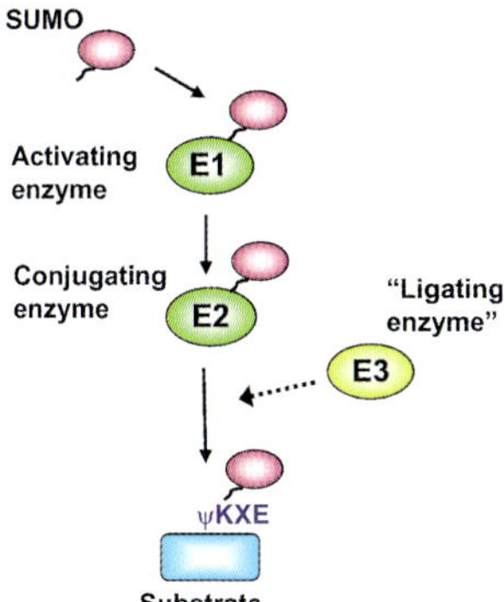

Fig. 1. The SUMO pathway. The E1 (SAE1/2), E2 (Ubc9) and E3 ligating enzymes are shown. The E3 proteins act to facilitate SUMO conjugation onto substrate proteins through lysine residues in the ψKxE motif, where ψ is a hydrophobic amino acid and x is any amino acid

These are thought to act either as adaptors that recruit the substrate to Ubc9 or in some cases to possess intrinsic enzymatic activity that promotes substrate sumoylation. There are three distinct classes of E3 enzymes that have been identified to date. The PIAS proteins possess a RING finger and are thought to act as adapter proteins (reviewed in Schmidt and Muller 2003). In contrast, Pc2 has been shown to possess both adapter and catalytic activity localized in distinct domains (Kagey et al. 2005). Similarly, Ran-BP is thought to possess catalytic activity and is thought to allosterically enhance the activity of Ubc9 (Pichler et al. 2004; Tatham et al. 2005). It is possible that other distinct classes of E3 ligase exist. Indeed, this would help to provide substrate specificity as observed for E3 ligases in the ubiquitin pathway and hence a possible route for specific regulation of substrate sumoylation.

11.3 Specificity and Regulation of Sumoylation

Substrate sumoylation usually takes place on lysine residues within motifs conforming to the consensus sequence ψKxE. Thus, at the substrate level, some specificity is imparted by the local context of the sumoylated lysine residue. It is possible that further specificity determinants exist. For example, two extended motifs have been identified that are common to subsets of substrate proteins which contain other conserved residues in addition to this core consensus. The SC (synergy control) motif contains several flanking proline residues (Subramanian et al. 2003), whereas in the R-motif several acidic residues are located downstream from the sumoylation sites (Yang et al. 2002). However, it is currently unclear what role these extended motifs play, although possible roles could be in providing extra specificity to substrate sumoylation and/or imparting context-dependent functional outcomes from sumoylation.

It is clear that sumoylation of many proteins causes an increase in their transcriptional repressive properties. However, in other cases such as the coactivator GRIP, sumoylation causes an increase in transcriptional activation (Kotaja et al. 2002a). Thus, functionally, specificity-determining mechanisms must exist. Indeed, SUMO itself has been shown to possess repressive properties (Ross et al. 2002; Yang et al. 2003a); therefore the output mediated by sumoylation must be modified in a context-dependent manner to permit nonrepressive responses.

A further important aspect of sumoylation is that it must be regulated to achieve changes in transcription factor activity status. Constitutive sumoylation would in many cases lead to constitutive transcriptional repression. However, regulatory mechanisms are likely to exist to both establish the sumoylated state and reverse this in response the appropriate cellular cues. Indeed, several transcriptional regulatory proteins change their sumoylation status in response to cellular stress (Fig. 2). For example, enhancement of CREB sumoylation is observed following stress induced by prolonged hypoxia (Comerford et al. 2003). Similarly, heat shock has been shown to enhance sumoylation of HSF1. However, in this case, phosphorylation is also induced and this phosphorylation is required to permit enhanced sumoylation to occur (Hietakangas et al. 2003). Further links with phosphorylation are provided by the regulation of PML by ERK MAP kinase in response to arsenic trioxide treatment. ERK-mediated PML phosphorylation is essential for the increases in sumoylation observed under these conditions (Hayakawa and Privalsky 2004). A further key example of regulated sumoylation is on the transcription factor Elk-1. However, in this case, SUMO is removed upon

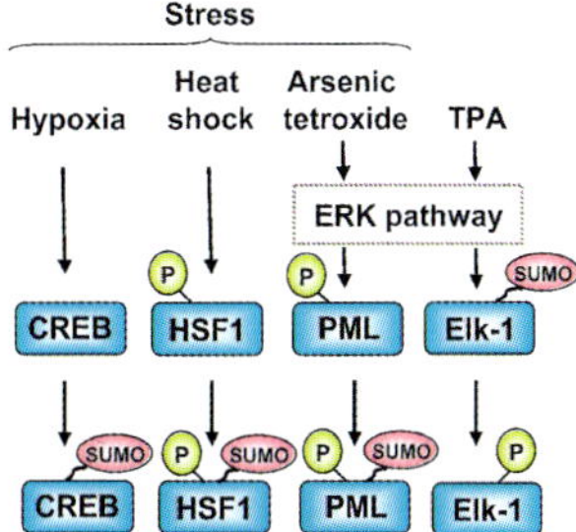

Fig. 2. Regulated substrate modification. Transcriptional regulatory proteins are shown whose sumoylation status changes in response to cellular stimuli. Substrate phosphorylation (*P*) occurring in response to each stimulus is depicted by a *yellow circle*. CREB, HSF1 and PML show enhanced sumoylation in response to different stress signals, whereas mitogenic signalling through the ERK pathway promotes reduced Elk-1 sumoylation

activation of the ERK MAP kinase pathway in response to mitogenic signals (Yang et al. 2003a).

11.4 Elk-1, MAP Kinase Pathways and Regulated Sumoylation

Mammalian cells contain several MAP kinase signalling cascades (reviewed in Roux and Blenis 2004). These can be grouped into the stress inducible pathways, including the p38 and JNK pathways, and mitogenic pathways exemplified by the ERK pathway. The pathways are activated through cell surface receptors, and the resulting signal is transduced by a series of GTPases, adapter proteins and protein kinases to the terminal MAP kinases ERK, JNK and p38. Upon activation, these protein kinases translocate to the nucleus where they impact on gene regulation by either directly phosphorylating transcriptional regulatory proteins or by activating downstream kinases, which themselves impact on transcriptional regulation (reviewed in Yang et al. 2003b). The ETS-domain transcription factor Elk-1 is a well characterized target of the MAP kinase pathways, and phosphorylation of its transcriptional activation domain (TAD) causes an overall enhancement of the transcriptional activation activity of Elk-1 (reviewed in Sharrocks 2002; Shaw and Saxton 2003). This activation is mediated by phosphorylation-dependent recruitment of the coactivator Sur2/Med23 (Stevens et al. 2002) and activation of the histone acetyl transferase (HAT) p300 (Li et al. 2003).

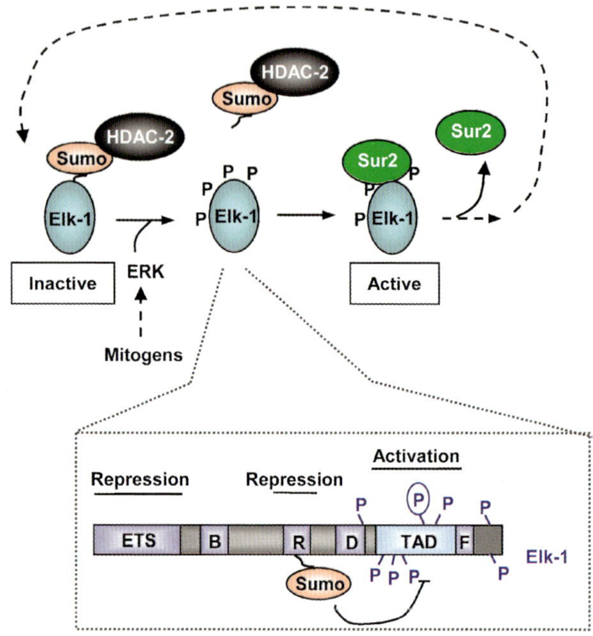

Fig. 3. ERK pathway regulation of Elk-1 sumoylation. Elk-1 exists in an inactive sumoylated state, and recruits HDAC-2. Activation of the ERK pathway by mitogenic signals leads to SUMO and HDAC-2 loss, Elk-1 phosphorylation and recruitment of the co-activator Sur2/MED23, thereby converting Elk-1 into an active state. The *inset* shows the domain structure of Elk-1 and associated transcriptional repressive and activating activities. The D and F domains are involved in MAP kinase recruitment. The location of MAP kinase phosphorylation sites (*P*) are shown, with the critical Ser383 within the TAD circled. The B-box is involved in promoter recruitment by binding to the transcription factor SRF. Elk-1 is sumoylated within the repressive R-motif and this inhibits the activity of the TAD

In addition to its C-terminal TAD, Elk-1 also contains two transcriptional repression domains (Fig. 3). One of these maps to the N-terminal ETS DNA-binding domain (Yang et al. 2001). The second motif, the R-motif, precedes the TAD and contains several SUMO modification sites (Yang et al. 2002, 2003a; Salinas et al. 2004). Sumoylation of Elk-1 within the R-motif is required for both its repressive properties (Yang et al. 2003a) and to reduce the rate of Elk-1 nucleocytoplasmic shuttling (Salinas et al. 2004). Importantly, this sumoylation is not a constitutive modification and is regulated in response to ERK MAP kinase signalling. Upon activation of the ERK pathway, Elk-1 sumoylation is lost and hence derepression occurs (Fig. 3) (Yang et al. 2003a). This occurs concomitantly with the increases in transcriptional activation mediated by enhanced co-activator recruitment/activation, leading to synergistic increases in Elk-1 transactivation potential.

11.5 Cross-talk Between Sumoylation and Histone Deacetylases

While it is well established that transcription factor sumoylation often correlates functionally with enhanced repressive properties, mechanistically, we are just beginning to uncover the molecular details of how this repression is elicited. One major route appears to be through SUMO-dependent recruitment of histone deacetylases (HDACs) (Fig. 4A). However, there appears to be specificity in the type of HDAC recruited as Elk-1 recruits HDAC-2, whereas p300 recruits HDAC-6 and histone H3 recruits HDAC-1 in a SUMO-dependent manner (Girdwood et al. 2003; Yang et al. 2004; Shiio and Eisenman 2003). It is currently not clear how this specificity is generated but presumably reflects the local context of the SUMO conjugation motif. Upon recruitment, the HDACs cause localized histone deacetylation, which itself correlates with repression of target genes. Indeed, it is often the balance between histone acetylase (HAT) and HDAC activities associated with a transcription factor that determines the transcriptional outcome on the promoter (Legube and Trouche 2003). This is illustrated in the case of Elk-1 (Fig. 4B), where in addition to triggering SUMO and HDAC-2 loss (Yang et al. 2004), activation of the ERK pathway enhances the activity of the HAT p300,

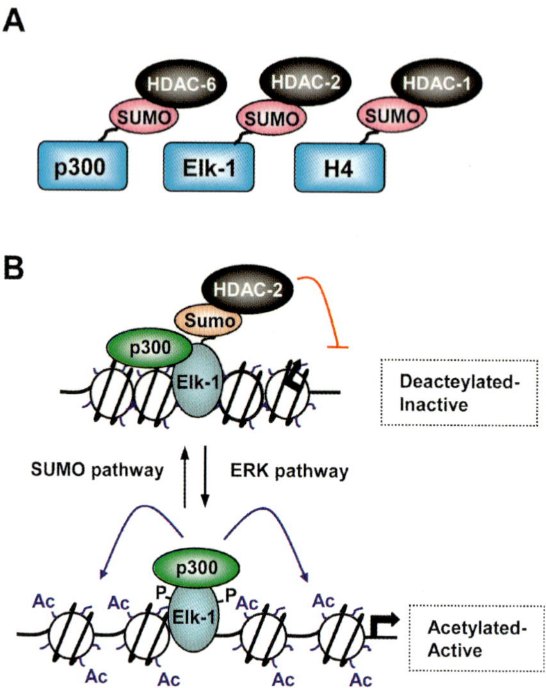

Fig. 4A,B. Interplay between histone deacetylases/acetylases and the SUMO pathway. **A** SUMO-dependent recruitment of specific HDACs by p300, Elk-1 and histone H4. **B** Changes in acetylase activity associated with Elk-1. Elk-1 sumoylation promotes HDAC-2 recruitment and hence local histone deacetylation and target gene inactivation. Activation of the ERK pathway causes HDAC-2 loss and concomitant activation of preassembled p300, and hence increased local histone acetylation (*Ac*) and target gene activation

which is constitutively associated with Elk-1 (Li et al. 2003). Together, these changes cause a net increase in localized histone acetylation that correlates with the activation of target genes. Thus, in the case of Elk-1, the ERK and SUMO pathways antagonize each other in regulating its transcriptional activity through changing the HDAC-HAT activity equilibrium and hence localized histone acetylation patterns. In addition, sumoylation can reverse interactions with HDACs. This has been shown for the DNA methyl transferase Dnmt3a, where sumoylation leads to the loss of HDAC-1 and -2 binding (Ling et al., 2004).

11.6 SUMO E3 Ligases

Several SUMO E3 ligases have been identified that can be grouped into three distinct classes by sequence similarity. These include Pc2 (Kagey et al. 2003), RanBP2 (Kirsh et al. 2002) and members of the PIAS family (reviewed in Schmidt and Muller 2003). The PIAS family all contain three conserved domains including a RING finger motif (Fig. 5A) and includes PIAS1, PIAS3, PIASy, PIASxα/β and hZimp10 (reviewed in Schmidt and Muller 2003). Analogously to the ubiquitin ligases, it is the RING finger motif that imparts E3 ligase activity on the PIAS proteins and mediates their function as adapter proteins (Kahyo et al. 2001; Kotaja et al. 2002b; Schmidt and Muller 2002). PIAS proteins also contain a SAP domain (reviewed in Aravind and Koonin 2000) and a small motif required for noncovalent binding to SUMO (Kotaja et al. 2002b; Song et al. 2004).

PIAS proteins have been shown to regulate a number of repressive and activating activities of transcription factors (Fig. 5B). The function of E3 ligases is thought to be to enhance the sumoylation of target proteins through acting at least in part as adapter proteins which permit enhanced substrate-Ubc9 interactions. Thus, by having many different E3 proteins, substrate specificity in the sumoylation reaction could be achieved. Indeed, enhanced SUMO modification mediated by PIAS proteins has been shown to be functionally important in enhancing the repressive activities of several transcription factors as observed for PIASxβ on p53 (Schmidt and Muller 2002). Furthermore, PIAS proteins were first identified by their ability to inhibit the transcriptional activity of STAT proteins (Chung et al. 1997) and this repression needs the sumoylation activity of the PIAS proteins (Rogers et al. 2003; Ungureanu et al. 2003).

PIAS proteins can also function to enhance the transcriptional activation capacity of target proteins as exemplified by the effects of PIASy on Tcf-4 (Yamamoto et al. 2003) and Zimp10 on the androgen receptor (Sharma et al. 2003). Again, in both cases, activation is thought to be

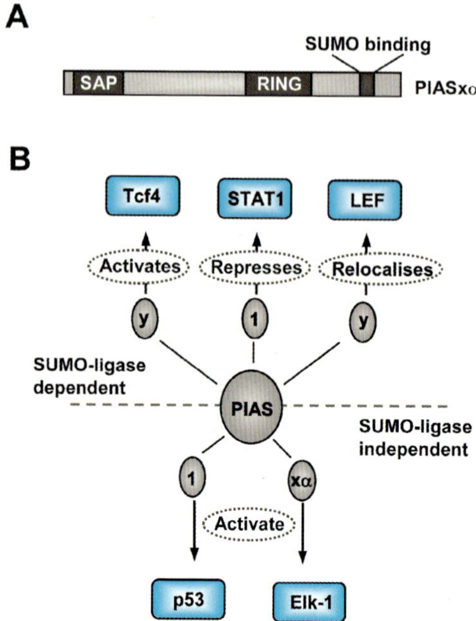

Fig. 5A,B. PIAS proteins and transcription factor regulation. **A** Domain structure of PIASxα. The domains shown are conserved in all PIAS proteins. **B** Different roles of PIAS proteins in controlling transcriptional factor activity. The different PIAS proteins are indicated by *small grey circles*, and target transcription factors by *blue rectangles*

due to the SUMO ligase activity of the PIAS proteins. In addition, PIAS proteins have been shown to affect transcription factor function through altering their subnuclear localization as shown for the action of PIASy on LEF1 (Sachdev et al. 2001). The sumoylation activity of PIASy is again important for this process, although the SUMO acceptor sites on LEF1 are not required.

However, the SUMO ligase activity is not always required and p53 is activated by PIAS1 in the absence of its RING finger motif (Megidish et al. 2002). Recent studies on Elk-1 have shown that PIASxα acts as a co-activator protein and this activity does not require its E3 ligase

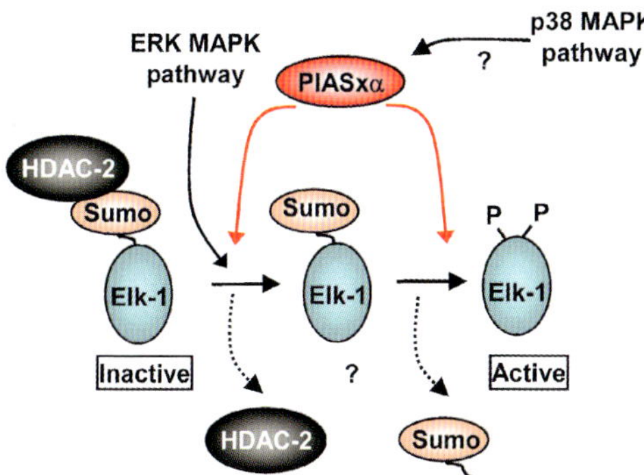

Fig. 6. PIASxα as an Elk-1 coactivator. PIASxα is required for the loss of SUMO and HDAC-2 from Elk-1 upon stimulation of the ERK MAP kinase pathway. It is currently unclear whether this occurs in one or two steps (proposed intermediary sumoylated form is indicated by a *question mark*). PIASxα can be phosphorylated by p38 MAP kinase, but it is unclear how this affects PIASxα-mediated regulation of Elk-1 (indicated by a *question mark*)

activity (Yang and Sharrocks 2005). Interestingly though, co-activation by PIASxα needs the SUMO acceptor sites in Elk-1, implying that the sumoylation status is important. This appears counterintuitive as sumoylation of Elk-1 is associated with transcriptional repression. However, the solution to this apparent paradox is revealed by the observation that PIASxα functions to facilitate SUMO and HDAC-2 loss from Elk-1 in response to activation of the ERK pathway, and hence contribute to its derepression (Fig. 6). PIASxα therefore appears to be an integral part of the activation pathway that converts Elk-1 from a transcriptional repressor to a transcriptional activator protein.

11.7 Signalling to PIAS Proteins

Little is currently known about how the activity of PIAS proteins is regulated. However, this is an attractive point for controlling transcription factor sumoylation in a substrate-specific manner due to the large number of PIAS-like proteins and other E3 ligases. In contrast, there appears to be only one E1 and one E2 enzyme in mammalian cells, and thus changing the activities of these enzymes would lead to pleiotropic changes in substrate sumoylation rather than a specific response. Indeed, this is illustrated by the action of the adenoviral protein Gam1 that inhibits the E1 enzymes, thereby causing pleiotropic changes in sumoylation and gene expression (Boggio et al. 2004).

The activity of PIASxβ towards Smad transcription factors is thought to be regulated by the p38 MAP kinase pathway. PIASxβ has been shown to up-regulate Smad activity through enhancing substrate sumoylation (Ohshima and Shimotohno 2003). Smad sumoylation is further augmented by activation of the p38 pathway. Activation of the p38 pathway causes up-regulation of PIASxβ expression at the mRNA level and also appears to promote its stabilization at the protein level. Thus, through acting on PIASxβ, the p38 pathway can up-regulate the sumoylation levels and activity of Smad transcription factors. Currently, the specificity of this response is unknown and it is unclear whether other PIASxβ substrates are affected.

PIASx exists in two common splice forms, xβ and xα, which differ in their C-terminal regions. Recent studies on PIASxα demonstrate that it is a direct target for the p38 MAP kinase pathway, and that the phosphorylated residues lie within a region conserved between the xα and xβ isoforms (Yang and Sharrocks, unpublished data). However, it is currently unclear whether these phosphorylation events affect protein stability and how p38-mediated phosphorylation of PIASxα might affect its activity towards partner proteins such as Elk-1. One attractive possibility is that stress-mediated activation of the p38 pathway might cause a different outcome on Elk-1 regulated genes to that caused by stimulation with mitogenic agents. Indeed, to date, no ERK MAP kinase-mediated phosphorylation events have been identified on PIASxα, although enhanced protein levels are detectable in response to mitogenic stimulation (Yang and Sharrocks, unpublished data).

11.8 Discussion and Perspectives

It is now clear that sumoylation plays a major role in controlling the activities of a large number of transcriptional regulatory proteins and hence has a large impact on cellular mRNA transcriptional profiles. In particular, sumoylation has been shown to have a major role in imparting repressive properties on transcription factors, at least in part, through promoting HDAC recruitment. However, future work is needed to understand the detailed molecular mechanisms of how SUMO-mediated transcriptional repression occurs.

Another major area for future focus is in how substrate sumoylation is controlled. Elk-1 makes an attractive model for studying this facet of sumoylation as activation of the ERK MAP kinase signalling pathway leads to its desumoylation. PIASxα plays an important role in this desumoylation process, but how this is achieved is currently unknown. PIASxα is itself a target of the p38 MAP kinase pathway, providing another potential point of regulation. It is currently unknown how this might impact on the activity of PIASxα, but potentially could affect its sumoylation activity or indeed its co-activating activity towards substrates such as Elk-1. Further studies are needed to see if other PIAS proteins or more divergent classes of E3 ligases are targets for the MAP kinase or other signalling pathways.

The increasing number of E3 enzymes for the SUMO pathway being discovered provides further possibilities for specific substrate regulation. By analogy with the ubiquitin pathway, it is likely that substrate specificity is mediated by the E3 proteins. In terms of potential therapeutic strategies, these E3 proteins would therefore represent attractive targets to obtain a more specific response than targeting E1 and E2 enzymes which would cause pleiotropic affects on cellular substrate sumoylation. Indeed, many viruses and bacterial pathogens encode proteins that can cause global effects on cellular sumoylation status (reviewed in Wilson and Rangasamy et al. 2001). A good example of this is the adenoviral protein Gam1 that binds and inactivates the E1 enzyme SAE1/SAE2 and promotes its degradation (Boggio et al. 2004). This leads to reduced global levels of sumoylation and is required for providing an environment that promotes viral replication. As cancer cells often undergo similar molecular changes to those induced by viruses, it is possible that global sumoylation defects may also occur in cancer cells, although to date, no such links have been made. In these cases, therapeutic strategies would aim to restore normal levels of sumoylation by increasing the activity of the core components of the SUMO pathway. Defects in sumoylation of specific substrates may occur in cancer cells and impact on the tumorigenic pathways. This is hinted at by the observation that many SUMO targets are oncogenic or tumour suppressor proteins. Furthermore, several proteins that are involved in chromosomal translocations are SUMO targets such as PML in PML-RARα fusions. Treatment of acute promyelocytic leukaemia cells harbouring PML-RARα with arsenic trioxide leads to enhanced sumoylation of PML-RARα, suggesting that this might be part of the mechanism by which this agent leads to reversal of the leukaemic defects (Muller et al. 1998). Recent studies demonstrate that arsenic trioxide leads to changes in PML sumoylation through the ERK MAP kinase pathway (Hayakawa and Privalsky 2004). Thus, the MAP kinases may also represent an area for potential intervention which impacts on substrate sumoylation and subsequent tumorigenic consequences.

It is likely that future studies will identify more SUMO targets that are associated with cancer. More direct links with SUMO pathway defects may also emerge. Therapeutic intervention in the SUMO pathway may then provide an attractive potential route for curing or stopping the spread of cancer.

Acknowledgements. We thank James Witty for comments on the manuscript. The work in our laboratory was supported by the Wellcome Trust.

References

Aravind L, Koonin EV (2000) SAP – a putative DNA-binding motif involved in chromosomal organization. Trends Biochem Sci 25:112–114

Boggio R, Colombo R, Hay RT, Draetta GF, Chiocca S (2004) A mechanism for inhibiting the SUMO pathway. Mol Cell 16:549–561

Chung CD, Liao J, Liu B, Rao X, Jay P, Berta P, Shuai K (1997) Specific inhibition of Stat3 signal transduction by PIAS3. Science 278:1803–1805

Comerford KM, Leonard MO, Karhausen J, Carey R, Colgan SP, Taylor CT (2003) Small ubiquitin-related modifier-1 modification mediates resolution of CREB-dependent responses to hypoxia. Proc Natl Acad Sci U S A 100:986–991

Gill G (2003) Post-translational modification by the small ubiquitin-related modifier SUMO has big effects on transcription factor activity. Curr Opin Genet Dev 13:108–113

Gill G (2004) SUMO and ubiquitin in the nucleus: different functions, similar mechanisms? Genes Dev 18:2046–2059

Girdwood D, Bumpass D, Vaughan OA, Thain A, Anderson LA, Snowden AW, Garcia-Wilson E, Perkins ND, Hay RT (2003) p300 transcriptional repression is mediated by SUMO modification. Mol Cell 11:1043–1054

Hay RT (2001) Protein modification by SUMO. Trends Biochem Sci 26:332–333

Hayakawa F, Privalsky ML (2004) Phosphorylation of PML by mitogen-activated protein kinases plays a key role in arsenic trioxide-mediated apoptosis. Cancer Cell 5:389–401

Hietakangas V, Ahlskog JK, Jakobsson AM, Hellesuo M, Sahlberg NM, Holmberg CI, Mikhailov A, Palvimo JJ, Pirkkala L, Sistonen L (2003) Phosphorylation of serine 303 is a prerequisite for the stress-inducible SUMO modification of heat shock factor 1. Mol Cell Biol 23:2953–2968

Kagey MH, Melhuish TA, Wotton D (2003) The polycomb protein Pc2 is a SUMO E3. Cell 113:127–137

Kagey MH, Melhuish TA, Powers SE, Wotton D (2005) Multiple activities contribute to Pc2 E3 function. EMBO J 14:108–119

Kahyo T, Nishida T, Yasuda H (2001) Involvement of PIAS1 in the sumoylation of tumor suppressor p53. Mol Cell 8:713–718

Kirsh O, Seeler JS, Pichler A, Gast A, Muller S, Miska E, Mathieu M, Harel-Bellan A, Kouzarides T, Melchior F, Dejean A (2002) The SUMO E3 ligase RanBP2 promotes modification of the HDAC4 deacetylase. EMBO J 21:2682–2691

Kotaja N, Karvonen U, Janne OA, Palvimo JJ (2002a) The nuclear receptor interaction domain of GRIP1 is modulated by covalent attachment of SUMO-1. J Biol Chem 277:30283–30288

Kotaja N, Karvonen U, Janne OA, Palvimo JJ (2002b) PIAS proteins modulate transcription factors by functioning as SUMO-1 ligases. Mol Cell Biol 22:5222–5234

Legube G, Trouche D (2003) Regulating histone acetyltransferases and deacetylases. EMBO Rep 4:944–947

Li Q-J, Yang S-H, Maeda Y, Sladek FM, Sharrocks AD, Martins-Green M (2003) MAP kinase phosphorylation-dependent activation of Elk-1 leads to activation of the coactivator p300. EMBO J 22:281–291

Ling Y, Sankpal UT, Robertson AK, McNally JG, Karpova T, Robertson KD (2004) Modification of de novo DNA methyltransferase 3a (Dnmt3a) by SUMO-1 modulates its interaction with histone deacetylases (HDACs) and its capacity to repress transcription. Nucleic Acids Res 32:598–610

Megidish T, Xu JH, Xu CW (2002) Activation of p53 by protein inhibitor of activated Stat1 (PIAS1). J Biol Chem 277:8255–8259

Muller S, Hoege C, Pyrowolakis G, Jentsch S (2001) SUMO, ubiquitin's mysterious cousin. Nat Rev Mol Cell Biol 2:202–210

Muller S, Matunis MJ, Dejean A (1998) Conjugation with the ubiquitin-related modifier SUMO-1 regulates the partitioning of PML within the nucleus. EMBO J 17:61–70

Ohshima T, Shimotohno K (2003) Transforming growth factor-beta-mediated signaling via the p38 MAP kinase pathway activates Smad-dependent transcription through SUMO-1 modification of Smad4. J Biol Chem 278:50833–50842

Pichler A, Knipscheer P, Saitoh H, Sixma TK, Melchior F (2004) The RanBP2 SUMO E3 ligase is neither. Nat Struct Mol Biol 11:984–991

Rogers RS, Horvath CM, Matunis MJ (2003) SUMO modification of STAT1 and its role in PIAS-mediated inhibition of gene activation. J Biol Chem 278:30091–30097

Ross S, Best JL, Zon LI, Gill G (2002) SUMO-1 modification represses Sp3 transcriptional activation and modulates its subnuclear localization. Mol Cell 10:831–842

Roux PP, Blenis J (2004) ERK and p38 MAPK-activated protein kinases: a family of protein kinases with diverse biological functions. Microbiol Mol Biol Rev 68:320–344

Sachdev S, Bruhn L, Sieber H, Pichler A, Melchior F, Grosschedl R (2001) PIASy, a nuclear matrix-associated SUMO E3 ligase, represses LEF1 activity by sequestration into nuclear bodies. Genes Dev 15:3088–3103

Salinas S, Briancon-Marjollet A, Bossis G, Lopez MA, Piechaczyk M, Jariel-Encontre I, Debant A, Hipskind RA (2004) SUMOylation regulates nucleo-cytoplasmic shuttling of Elk-1. J Cell Biol 165:767–773

Schmidt D, Muller S (2002) Members of the PIAS family act as SUMO ligases for c-Jun and p53 and repress p53 activity. Proc Natl Acad Sci U S A 99:2872–2877

Schmidt D, Muller S (2003) PIAS/SUMO: new partners in transcriptional regulation. Cell Mol Life Sci 60:2561–2574

Sharma M, Li X, Wang Y, Zarnegar M, Huang CY, Palvimo JJ, Lim B, Sun Z (2003) hZimp10 is an androgen receptor co-activator and forms a complex with SUMO-1 at replication foci. EMBO J 22:6101–6114

Sharrocks AD (2002) Complexities in ETS-domain transcription factor function and regulation; lessons from the TCF subfamily. Biochem Soc Trans 30:1–9

Shaw PE, Saxton J (2003) Ternary complex factors: prime nuclear targets for mitogen-activated protein kinases. Int J Biochem Cell Biol 35:1210–1226

Shiio Y, Eisenman RN (2003) Histone sumoylation is associated with transcriptional repression. Proc Natl Acad Sci U S A 100:13225–13230

Song J, Durrin LK, Wilkinson TA, Krontiris TG, Chen Y (2004) Identification of a SUMO-binding motif that recognizes SUMO-modified proteins. Proc Natl Acad Sci U S A 101:14373–14378

Stevens JL, Cantin GT, Wang G, Shevchenko A, Shevchenko A, Berk AJ (2002) Transcription control by E1A and MAP kinase pathway via Sur2 mediator subunit. Science 296:755–758

Subramanian L, Benson MD, Iniguez-Lluhi JA (2003) A synergy control motif within the attenuator domain of CCAAT/enhancer-binding protein alpha inhibits transcriptional synergy through its PIASy-enhanced modification by SUMO-1 or SUMO-3. J Biol Chem 278:9134–9141

Tatham MH, Kim S, Jaffray E, Song J, Chen Y, Hay RT (2005) Unique binding interactions among Ubc9, SUMO and RanBP2 reveal a mechanism for SUMO paralog selection. Nat Struct Mol Biol 12:67–74

Ungureanu D, Vanhatupa S, Kotaja N, Yang J, Aittomaki S, Janne OA, Palvimo JJ, Silvennoinen O (2003) PIAS proteins promote SUMO-1 conjugation to STAT1. Blood 102:3311–3313

Verger A, Perdomo J, Crossley M (2003) Modification with SUMO: a role in transcriptional regulation. EMBO Rep 4:137–142

Wilson VG, Rangasamy D (2001) Viral interaction with the host cell sumoylation system. Virus Res 81:17–27

Yamamoto H, Ihara M, Matsuura Y, Kikuchi A (2003) Sumoylation is involved in beta-catenin-dependent activation of Tcf-4. EMBO J 22:2047–2059

Yang SH, Vickers E, Brehm A, Kouzarides T, Sharrocks AD (2001) Temporal recruitment of the mSin3A-histone deacetylase corepressor complex to the ETS domain transcription factor Elk-1. Mol Cell Biol 21:2802–2814

Yang SH, Bumpass DC, Perkins ND, Sharrocks AD (2002) The ETS domain transcription factor Elk-1 contains a novel class of repression domain. Mol Cell Biol 22:5036–5046

Yang SH, Jaffray E, Hay RT, Sharrocks AD (2003a) Dynamic interplay of the SUMO and ERK pathways in regulating Elk-1 transcriptional activity. Mol Cell 12:63–74

Yang SH, Sharrocks AD, Whitmarsh AJ (2003b) Transcriptional regulation by the MAP kinase signaling cascades. Gene 320:3–21

Yang SH, Sharrocks AD (2004) SUMO promotes HDAC-mediated transcriptional repression. Mol Cell 13:611–617

Yang SH, Sharrocks AD (2005) PIASx acts as an Elk-1 coactivator by facilitating derepression. EMBO J 24:2161–2171

Ernst Schering Research Foundation Workshop

Editors: Günter Stock
Monika Lessl

Vol. 1 (1991): Bioscience ⇌ Societly Workshop Report
Editors: D.J. Roy, B.E. Wynne, R.W. Old

Vol. 2 (1991): Round Table Discussion on Bioscience ⇌ Society
Editor: J.J. Cherfas

Vol. 3 (1991): Excitatory Amino Acids and Second Messenger Systems
Editors: V.I. Teichberg, L. Turski

Vol. 4 (1992): Spermatogenesis – Fertilization – Contraception
Editors: E. Nieschlag, U.-F. Habenicht

Vol. 5 (1992): Sex Steroids and the Cardiovascular System
Editors: P. Ramwell, G. Rubanyi, E. Schillinger

Vol. 6 (1993): Transgenic Animals as Model Systems for Human Diseases
Editors: E.F. Wagner, F. Theuring

Vol. 7 (1993): Basic Mechanisms Controlling Term and Preterm Birth
Editors: K. Chwalisz, R.E. Garfield

Vol. 8 (1994): Health Care 2010
Editors: C. Bezold, K. Knabner

Vol. 9 (1994): Sex Steroids and Bone
Editors: R. Ziegler, J. Pfeilschifter, M. Bräutigam

Vol. 10 (1994): Nongenotoxic Carcinogenesis
Editors: A. Cockburn, L. Smith

Vol. 11 (1994): Cell Culture in Pharmaceutical Research
Editors: N.E. Fusenig, H. Graf

Vol. 12 (1994): Interactions Between Adjuvants, Agrochemical and Target Organisms
Editors: P.J. Holloway, R.T. Rees, D. Stock

Vol. 13 (1994): Assessment of the Use of Single Cytochrome P450 Enzymes in Drug Research
Editors: M.R. Waterman, M. Hildebrand

Vol. 14 (1995): Apoptosis in Hormone-Dependent Cancers
Editors: M. Tenniswood, H. Michna

Vol. 15 (1995): Computer Aided Drug Design in Industrial Research
Editors: E.C. Herrmann, R. Franke

Vol. 16 (1995): Organ-Selective Actions of Steroid Hormones
Editors: D.T. Baird, G. Schütz, R. Krattenmacher

Vol. 17 (1996): Alzheimer's Disease
Editors: J.D. Turner, K. Beyreuther, F. Theuring

Vol. 18 (1997): The Endometrium as a Target for Contraception
Editors: H.M. Beier, M.J.K. Harper, K. Chwalisz

Vol. 19 (1997): EGF Receptor in Tumor Growth and Progression
Editors: R.B. Lichtner, R.N. Harkins

Vol. 20 (1997): Cellular Therapy
Editors: H. Wekerle, H. Graf, J.D. Turner

Vol. 21 (1997): Nitric Oxide, Cytochromes P 450,
and Sexual Steroid Hormones
Editors: J.R. Lancaster, J.F. Parkinson

Vol. 22 (1997): Impact of Molecular Biology
and New Technical Developments in Diagnostic Imaging
Editors: W. Semmler, M. Schwaiger

Vol. 23 (1998): Excitatory Amino Acids
Editors: P.H. Seeburg, I. Bresink, L. Turski

Vol. 24 (1998): Molecular Basis of Sex Hormone Receptor Function
Editors: H. Gronemeyer, U. Fuhrmann, K. Parczyk

Vol. 25 (1998): Novel Approaches to Treatment of Osteoporosis
Editors: R.G.G. Russell, T.M. Skerry, U. Kollenkirchen

Vol. 26 (1998): Recent Trends in Molecular Recognition
Editors: F. Diederich, H. Künzer

Vol. 27 (1998): Gene Therapy
Editors: R.E. Sobol, K.J. Scanlon, E. Nestaas, T. Strohmeyer

Vol. 28 (1999): Therapeutic Angiogenesis
Editors: J.A. Dormandy, W.P. Dole, G.M. Rubanyi

Vol. 29 (2000): Of Fish, Fly, Worm and Man
Editors: C. Nüsslein-Volhard, J. Krätzschmar

Vol. 30 (2000): Therapeutic Vaccination Therapy
Editors: P. Walden, W. Sterry, H. Hennekes

Vol. 31 (2000): Advances in Eicosanoid Research
Editors: C.N. Serhan, H.D. Perez

Vol. 32 (2000): The Role of Natural Products in Drug Discovery
Editors: J. Mulzer, R. Bohlmann

Vol. 33 (2001): Stem Cells from Cord Blood, In Utero Stem Cell Development, and Transplantation-Inclusive Gene Therapy
Editors: W. Holzgreve, M. Lessl

Vol. 34 (2001): Data Mining in Structural Biology
Editors: I. Schlichting, U. Egner

Vol. 35 (2002): Stem Cell Transplantation and Tissue Engineering
Editors: A. Haverich, H. Graf

Vol. 36 (2002): The Human Genome
Editors: A. Rosenthal, L. Vakalopoulou

Vol. 37 (2002): Pharmacokinetic Challenges in Drug Discovery
Editors: O. Pelkonen, A. Baumann, A. Reichel

Vol. 38 (2002): Bioinformatics and Genome Analysis
Editors: H.-W. Mewes, B. Weiss, H. Seidel

Vol. 39 (2002): Neuroinflammation – From Bench to Bedside
Editors: H. Kettenmann, G.A. Burton, U. Moenning

Vol. 40 (2002): Recent Advances in Glucocorticoid Receptor Action
Editors: A. Cato, H. Schaecke, K. Asadullah

Vol. 41 (2002): The Future of the Oocyte
Editors: J. Eppig, C. Hegele-Hartung

Vol. 42 (2003): Small Molecule-Protein Interaction
Editors: H. Waldmann, M. Koppitz

Vol. 43 (2003): Human Gene Therapy:
Present Opportunities and Future Trends
Editors: G.M. Rubanyi, S. Ylä-Herttuala

Vol. 44 (2004): Leucocyte Trafficking:
The Role of Fucosyltransferases and Selectins
Editors: A. Hamann, K. Asadullah, A. Schottelius

Vol. 45 (2004): Chemokine Roles in Immunoregulation and Disease
Editors: P.M. Murphy, R. Horuk

Vol. 46 (2004): New Molecular Mechanisms of Estrogen Action and Their Impact on Future Perspectives in Estrogen Therapy
Editors: K.S. Korach, A. Hillisch, K.H. Fritzemeier

Vol. 47 (2004): Neuroinflammation in Stroke
Editors: U. Dirnagl, B. Elger

Vol. 48 (2004): From Morphological Imaging to Molecular Targeting
Editors: M. Schwaiger, L. Dinkelborg, H. Schweinfurth

Vol. 49 (2004): Molecular Imaging
Editors: A.A. Bogdanov, K. Licha

Vol. 50 (2005): Animal Models of T Cell-Mediated Skin Diseases
Editors: T. Zollner, H. Renz, K. Asadullah

Vol. 51 (2005): Biocombinatorial Approaches for Drug Finding
Editors: W. Wohlleben, T. Spellig, B. Müller-Tiemann

Vol. 52 (2005): New Mechanisms for Tissue-Selective Estrogen-Free Contraception
Editors: H.B. Croxatto, R. Schürmann, U. Fuhrmann, I. Schellschmidt

Vol. 53 (2005): Opportunities and Challenges of the Therapies Targeting CNS Regeneration
Editors: D. Perez, B. Mitrovic, A. Baron Van Evercooren

Vol. 54 (2005): The Promises and Challenges of Regenerative Medicine
Editors: J. Morser, S.I. Nishikawa

Vol. 55 (2006): Chronic Viral and Inflammatory Cardiomyopathy
Editors: H.-P. Schultheiss, J.-F. Kapp

Vol. 56 (2006): Cytokines as Potential Therapeutic Target for Inflammatory Skin Diseases
Editors: R. Numerof, C.A. Dinarello, K. Asadullah

Vol. 57 (2006): The Histone Code and Beyond
Editors: S.L. Berger, O. Nakanishi, B. Haendler

Supplement 1 (1994): Molecular and Cellular Endocrinology of the Testis
Editors: G. Verhoeven, U.-F. Habenicht

Supplement 2 (1997): Signal Transduction in Testicular Cells
Editors: V. Hansson, F.O. Levy, K. Taskén

Supplement 3 (1998): Testicular Function:
From Gene Expression to Genetic Manipulation
Editors: M. Stefanini, C. Boitani, M. Galdieri, R. Geremia,F. Palombi

Supplement 4 (2000): Hormone Replacement Therapy
and Osteoporosis
Editors: J. Kato, H. Minaguchi, Y. Nishino

Supplement 5 (1999): Interferon: The Dawn of Recombinant
Protein Drugs
Editors: J. Lindenmann, W.D. Schleuning

Supplement 6 (2000): Testis, Epididymis and Technologies
in the Year 2000
Editors: B. Jégou, C. Pineau, J. Saez

Supplement 7 (2001): New Concepts in Pathology
and Treatment of Autoimmune Disorders
Editors: P. Pozzilli, C. Pozzilli, J.-F. Kapp

Supplement 8 (2001): New Pharmacological Approaches
to Reproductive Health and Healthy Ageing
Editors: W.-K. Raff, M.F. Fathalla, F. Saad

Supplement 9 (2002): Testicular Tangrams
Editors: F.F.G. Rommerts, K.J. Teerds

Supplement 10 (2002): Die Architektur des Lebens
Editors: G. Stock, M. Lessl

Supplement 11 (2005): Regenerative and Cell Therapy
Editors: A. Keating, K. Dicke, N. Gorin, R. Weber, H. Graf

Supplement 12 (2005): Von der Wahrnehmung zur Erkenntnis –
From Perception to Understanding
Editors: M. Lessl, J. Mittelstraß

This series will be available on request from
Ernst Schering Research Foundation, 13342 Berlin, Germany